Peirce's
Philosophy of Science

Peirce's Philosophy of Science

Critical Studies in His Theory of Induction and Scientific Method

NICHOLAS RESCHER

UNIVERSITY OF NOTRE DAME PRESS
NOTRE DAME — LONDON

Library of Congress Cataloging in Publication Data

Rescher, Nicholas.
Peirce's philosophy of science.

Bibliography: p.
Includes index.
1. Science—Philosophy. 2. Science—Methodology.
3. Peirce, Charles Santiago Sanders, 1839–1914.
I. Title.
Q175.R39333 501 77–82479
ISBN 0–268–01526–0
ISBN 0–268–01527–9 pbk.

Manufactured in the United States of America

For My Mother
with Admiration and Gratitude

Contents

Preface ix

1. Peirce's Theory of the Self-correctiveness of Science 1

2. Peirce on Scientific Progress and the Completability of Science 19

3. Peirce on Abduction, Plausibility, and the Efficiency of Scientific Inquiry 41

4. Peirce and the Economy of Research 65

Notes 93

References 117

Index of Names 121

Index of Subjects 123

Preface

My concern with Peirce is by no means strictly historical. I look on him not as a thinker of bygone days, but as a colleague and co-worker on issues of abiding interest. Time and again I have run up against him in the course of working out my own lines of thought.[1] I have come to regard Peirce as a more fertile and stimulating thinker than most in that somewhat scholastic tradition of the philosophy of science that has sprung up since his day. More than any other student of the nature of science, he pries into the things we always wanted to know but were afraid to ask.

The proximate cause of this book was an invitation to give a group of lectures in the Perspectives in Philosophy series at the University of Notre Dame during the 1976 fall term. The overall theme was to be American philosophy, as was only fitting in the bicentennial year. When this invitation was extended, I immediately fixed on Peirce as my subject and the present book is the outgrowth of these lectures.

I am indebted to Cynthia Freeland for reading a draft of the manuscript and suggesting some improvements. I owe thanks to Kathleen Reznik for preparing the typescript through a sequence of revisions and to Jay Garfield and Virginia Chestek for help with checking the proofs. I also want to express my gratitude to my philosophical col-

1. This becomes readily apparent from the name indices of three of my recent books: *The Primacy of Practice* (Oxford, 1973); *Methodological Pragmatism* (Oxford, 1977); and *Scientific Progress* (Oxford, 1977).

leagues at Notre Dame—faculty and students alike—for making my visit there a thoroughly pleasant experience.

Pittsburgh
February 1977

1 Peirce's Theory of the Self-correctiveness of Science

1. Peirce's Position and the Critical Onslaught upon It

Everyone realizes that we must turn to science to understand the ways of the world; to describe, explain, and predict the phenomena of nature. But how do we know that science is really efficacious and that the scientific method of inquiry is successful in getting at the real truth of things?

For Peirce, the inductive method used in the sciences leads inevitably to truth; its justification lies in being self-corrective. It has the capacity to yield the correct result in the long run, whatever transitory errors and missteps may occur along the way. Peirce saw self-correction as the definitive characteristic of induction, which in its very nature "is a method of reaching conclusions which, if persisted in long enough, will assuredly correct any error concerning future experience into which it may temporarily lead us."[1] Peirce put the matter forcefully in his essay, "The Logic of Drawing History from Ancient Documents":

> Induction . . . is not justified by any relation between the facts stated in the premisses and the fact stated in the conclusion; and it does not infer that the latter fact is either necessary or objectively probable. But the justification of its conclusion is that that conclusion is reached by *a method which, steadily persisted in, must lead to true knowledge in the long run of cases of its application, whether to the existing world or to any*

> *imaginable world whatsoever*. (CP, 7.207 [1901]; italics added)[2]

The operational adequacy of our inductive reasonings can be checked, and indeed improved, by inductive means. Induction is thus self-monitoring. It has the characteristic, crucial to its justification as a rational resource, that its continued use will in the long run uncover any mistakes to which its previous use may have led. Its nature is such that science, if persistent in its use, is predestined to reach the truth eventually. And thus, science—not science in our own day, but ultimate-long-range-science—is infallible.[3]

No part of Peirce's philosophy of science has been more severely criticized, even by his most sympathetic commentators, than this attempted validation of inductive methodology on the basis of its purported self-correctiveness. Even dedicated Peirceans, let along his critics, incline to meet his claims in this area with a mixture of incredulity and dismay. This book, however, will endeavor to rehabilitate this aspect of Peirce's theory of scientific method, arguing that his views on the inductive corrigibility of our inductive practices are both coherent and cogent.

Let us begin at the beginning and consider the structure of Peirce's theory of induction. In his Lowell Lectures of 1903, and in many other places, Peirce distinguished three modes of inductive reasoning: *crude* (or *rudimentary*) induction, *qualitative* induction, and *quantitative* (or *statistical*) induction.[4]

Crude induction is concerned with the projection of universal claims on the basis of a uniform experience of conforming instances. It relies on "the *absence* of [any encountered] instances to the contrary."[5] The merely empirical generalizations of everyday life—"All swans are white," "Thunder is always preceded by lightning"—are illustrations. Such generalizations simply dismiss the prospect that familiar things might fall outside familiar patterns. This crude sort of induction will not concern us much here; Peirce rightly views it as a primitive instrumentality of everyday life that plays little or no role in scientific inquiry.

Qualitative induction is regarded by Peirce as a powerful instrument of very general utility in inquiry. In its essentials it is simply equivalent to the *hypothetico-deductive method*. Phenomena are observed. A series of explanatory hypotheses—$H_1, H_2, \ldots, H_n$—is imaginatively projected to account for these (by the process of conjectural hypothesis-proliferation that Peirce called *abduction*). These hypotheses are then tested by the familiar process of exploiting them as a basis for predictions, which are then checked against the actual course of developments. The hypothesis that fares best under such trial is tentatively adopted over the alternatives until it is itself overthrown by a further sequence of projection and testing of hypotheses. Peirce gave the name *retroduction* to this process of eliminating hypotheses by experiential/experimental testing. Qualitative-induction is thus the collaborative meshing of abduction and retroduction, of hypothesis conjecture and hypothesis testing.

The third mode of induction, quantitative induction is in effect the methodology of *statistics*, the mathematically guided process of sampling and sample-analysis applied in the context of scientific reasoning. Its mode of operation is largely based on the precept that later writers on inductive inference have called "the straight rule" of induction, the principle that the observed frequency of some target-property in a sample may be taken (if the sampling process is appropriately designed) as an index for its actual frequency in the population at large.[6] In quantitative induction, *observed* frequencies are taken as indicators of *actual* frequencies.

The pivotal feature of the process of quantitative induction is, according to Peirce, that it is *automatically* self-monitoring or self-corrective. If the frequency with which some target-property is distributed over the individuals of a sample does not correspond to its frequency of distribution over the population, the discrepancy is certain to become apparent as the sampling process is extended over the long run. By constantly readjusting the estimate of the population-frequency in light of the actual sample-frequency, the scientist is bound eventually to get things right.

As Peirce sees it, quantitative induction is effective as a matter of abstract mathematical principle, independent of any factual presuppositions or dubious metaphysical assumptions:

> [An] endless series must have some character; and it would be absurd to say that experience has a character which is never manifested. But there is no other way in which the character of that series can manifest itself than while the endless series is still incomplete. Therefore, if the character manifested by the series up to a certain point is not that character which the entire series possesses, still, as the series goes on, it must eventually tend, however irregularly, towards becoming so; and all the rest of the reasoner's life will be a continuation of this inferential process. This inference does not depend upon any assumption that the series will be endless, or that the future will be like the past, or that nature is uniform, nor upon any material assumption whatever. (CP, 2.784 [c. 1905])

Quantitative induction, Peirce maintains, is unfailingly successful in the long run and "always makes a gradual approach to the truth, though not a uniform approach."[7]

The close analogy of Peirce's position here to Hans Reichenbach's well-known pragmatic justification of induction is striking (indeed, it eventually struck Reichenbach himself). Most modern writers on Peirce's theory of induction, influenced no doubt by Reichenbach's spirited defense of a closely similar position, are inclined to concede that Peirce's doctrine of the self-correctiveness of quantitative induction is a theory of, at any rate, some discussable merit. They view it not, to be sure, as decisively established, but at any rate as a controversial position for whose tenability (duly construed and qualified) a reasonable case can be made.[8]

Here the commentators reach a sticking point, however, for Peirce's own explicit and considered position is not merely that *quantitative* induction is self-corrective but that induction as it figures in scientific practice in general is self-corrective. In sum, induction at large is self-corrective. How, the critics ask, can Peirce possibly maintain this self-correctiveness, since scientific induction is preeminently based on *qualitative* induction? On what grounds can Peirce hold that the hypothetico-deductive method in general is self-corrective, even in the case of nonstatistical hypotheses? As G. H. von Wright puts it:

> The Peircean idea of induction as a self-correcting approximation of the truth has no immediate significance . . . for other types of inductive reasoning than statistical generalization [i.e., specifically quantitative induction]. (*The Logical Problem of Induction*, 2nd ed. [Oxford, 1965], p. 226)

Abner Shimony objects as follows:

> [T]he only clear example of an infallible asymptotic approach which he [Peirce] offers is the simple one which is at the heart of Reichenbach's treatment of scientific inference: the evaluation of the limit of relative frequencies in infinite sequences of events (for example 2.650, 6.100, 7.111, 7.120). Since this kind of inference (''statistical'' or ''quantitative'' induction) is only one of the three kinds of induction which he recognizes . . . even sympathetic commentators on Peirce have found that his demonstrations [of the self-correctiveness of science] fall far short of realizing his general program. (Shimony 1970, p. 127)

Laurens Laudan appears to speak for all of Peirce's critics here:

> [Peirce] says quite plainly that all forms of induction are self-corrective. . . . And I think it would be less than candid not to say that Peirce offers no cogent reasons, not even mildly convincing ones, for believing that most inductive methods [indeed, all apart from quantitative induction] are self-corrective. . . . Seemingly unwilling to admit, even to himself, that he has failed in his original intention to establish . . . [self-correction] for all the methods of science, Peirce acts as if his argument about quantitative induction shows all the other species of induction to be self-corrective as well. (Laudan 1973, p. 293)[9]

Virtually to a man, Peirce's expositors, even the more sympathetic, are prepared to desert him on this issue.[10]

Such complaints and objections set the stage for the present inquiry, whose focus is indicated by the following seemingly incongruous triad:

1. Peirce maintains that the inductive methodology of science is internally complex and specifically includes not only *quantitative* but also *qualitative* induction.
2. Peirce establishes (with at least some degree of cogency) only the self-correctiveness of quantitative induction.
3. Peirce maintains self-correctiveness to be a crucial and characteristic aspect of scientific methodology in general.

How can Peirce (or a sympathetic Peircean) possibly reconcile these three points? The commentators think that it cannot be done. The ensuing discussion will endeavor to show that a reconciliation is indeed possible.

2. An Interpretative Reconstruction of Peirce's Position

To begin with, let the cards be spread on the table, presenting an interpretative reconstruction of a Peircean position that straightforwardly reconciles the three seemingly discordant theses under consideration. This interpretation is based on a critical distinction. If a productive process P is composed of several constituent subprocesses P_1, P_2, etc., then the idea of self-corrective monitoring of performance can be construed in two ways.

1. *Distributively*. Each constitutive cognitive process P_i monitors *its own* performance: for every P_i belonging to P, P_i monitors the performance of *itself*.
2. *Collectively*. Some one (or, conceivably, several of the processes P_i monitors the performance *of P as a whole*, and this unit is in turn *self*-corrective on its own account; the overall performance of P is thus monitored by certain of its constituent components, which are themselves self-monitoring.

The difference can be illustrated graphically as follows.

1. *Distributive Self-correctiveness*

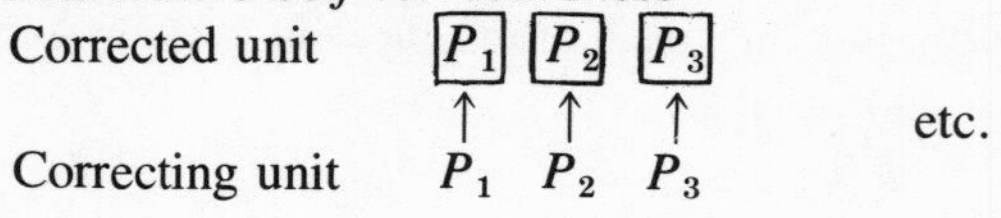

Corrected unit	$\boxed{P_1}$	$\boxed{P_2}$	$\boxed{P_3}$	
	$\uparrow$	$\uparrow$	$\uparrow$	etc.
Correcting unit	P_1	P_2	P_3	

2. *Collective Self-correctiveness*

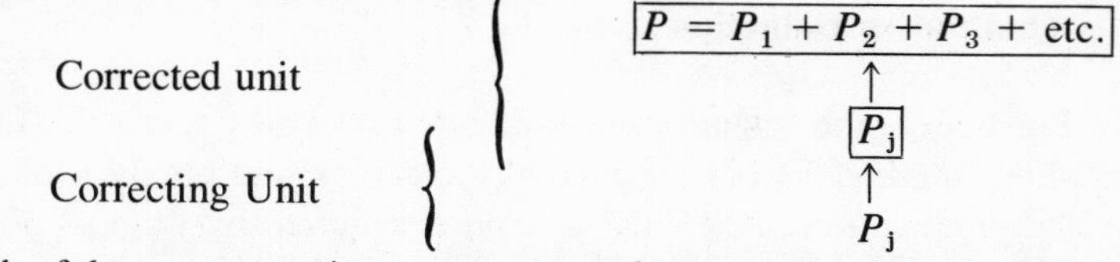

Both of these constructions represent perfectly good, although differing, modes of "self-correction."

Now the crucial fact is that the previously cited criticisms of Peirce's theory of the self-correctiveness of induction are all based on the specifically *distributive* construction of self-correction. The critics, in effect, complain to Peirce, "You say that scientific methodology has several components (preeminently qualitative and quantitative induction). But you only argue explicitly for the self-correctiveness of quantitative induction. How can you go on to claim that science as a whole is self-corrective?"

This complaint may be met on Peirce's behalf by pointing out that what is at issue in his theory of scientific method is the self-correctiveness of scientific induction in the *collective* sense, with quantitative induction in the controlling role. The crux of Peirce's view is that the performance of scientific induction as a whole, qualitative induction specifically included, can be monitored by statistical techniques, which in their turn are self-corrective in the narrower sense.

Exactly what must be established to support the view that scientific induction at large is self-corrective? Two theses are clearly needed:

1. Quantitative induction is (individually) self-corrective.
2. Qualitative induction can be correctively monitored by quantitative induction.

The first of these has already been discussed. It is certainly clear that Peirce does indeed maintain this thesis, and does so on grounds whose tenability—or at any rate arguability—is generally conceded. We may thus concentrate on the second thesis, the contention that the performance of qualitative induction can be monitored by statistical means.

3. How Can Statistical Methods Monitor the Performance of Qualitative Induction?

For Peirce, the scientific method is not a small list of rules (à la "Mill's Methods") but an internally complex and highly sophisticated organon, an intellectual discipline acquired by years of study and apprenticeship in the actual practice of theorizing and experimenting.

Peirce sees qualitative induction as an evolutionary process of variation and selection. Two component processes are involved here, as we have seen:

1. Hypothesis-projection or *abduction*: the purely conjectural proliferation of a whole gamut of alternative explanatory hypotheses that are relatively plausible, à proliferation based on guesswork—though not "mere" guesswork, but guesswork guided by a scientifically trained intuition. The aim of this enterprise is to identify those hypotheses that merit detailed scrutiny.
2. Hypothesis-testing or *retroduction*: the elimination of hypotheses on the basis of observational data, generally secured by suitably contrived experimental trials.

The result of the operation of the overall process is that the progress of science proceeds by the repeated eliminations of rival hypotheses in favor of one preferred candidate, each stage of the abduction-retroduction cycle reducing a cluster of conjectural hypotheses to an accepted theory. The systematic operation of the scientific method thus results at every stage in a diversified family of accepted hypotheses (that is, theories) that collectively constitute "the (current) scientific view of the way in which things work in the world."

The crucial feature of Peirce's theory of scientific methodology is that this process of theory and acceptance can be monitored statistically in terms of the *applications* of theories. As Peirce puts it, in a letter to Paul Carus, in qualitative induction the inquirer "may turn to the consideration of the hypothesis, study it thoroughly, and deduce miscellaneous observable consequences, and *then* return to the phenomena to find out how nearly these consequences agree with the actual facts'' (CP, 8.232 [c. 1910]). Each time we employ a theory for prediction or for actual control (using it to guide intervention in the course of natural events to produce a desired outcome), we contribute to the statistical sample population by which its credentials are monitored. Its adequacy is thus controlled by what might be called its ''success ratio'':

$$\frac{\text{Number of successful applications (in prediction or control)}}{\text{Total number of such applications.}}$$ [11]

To put the matter in Peirce's own words:

> [T]he only sound procedure for induction, whose business consists in testing a hypothesis . . . is to receive its suggestions from the hypotheses first, to take up the predictions of experience which it conditionally makes, and then try the experiment. . . . [W]hen we get to the inductive stage *what we are about is finding out how much like the truth our hypothesis is, that is, what proportion of its anticipations will be verified*. (CP, 2.755 [c. 1905], italics added)[12]

The effectiveness of this success-ratio as a control of theorizing about the world is a key aspect of Peirce's understanding of the very idea of physical *reality* itself:

> [T]he distinction of reality and fiction depends on the supposition that sufficient investigation would cause one opinion to be universally received and all others to be rejected. . . . [I]n the long run, there is a real fact which corresponds to the idea . . . that a given mode of inference sometimes proves successful and sometimes not, and that in a ratio ultimately fixed. As we go on drawing inference after inference of the given kind, during the first ten or hundred cases the ratio of successes may be expected to show considerably fluctuations; but when we come into the thousands and millions, these fluctuations become less and less; and if we continue long enough, the ratio will approximate toward a fixed limit. We may, therefore,

> define the probability of a mode of argument as the proportion of cases in which it carries truth with it. (CP, 2.650 [1878])

Thus a true theory is one whose success ratio throughout the range of its inferential applications is 100 percent, one whose probability of correctness is 1, one that applies effectively throughout the whole range of the contexts of its inferential applications.[13] And a theory will more closely approximate the truth as its success ratio is higher, that is, closer to 1. It is against this background that Peirce's validation of induction should be seen:

> Thus, an argument [i.e., a "probable deductive" argument] that out of a certain set of sixty throws of a pair of dice about to be thrown, about ten will probably be doublets, is rendered valid by the fact that if a great number of just such arguments were made, the immense majority of the conclusions would be true, and indeed ten would be indefinitely near the actual average number in the long run. The validity of induction is entirely different; for it is by no means certain that the conclusion actually drawn in any given case would turn out true in the majority of cases where precisely such a method was followed; but what is certain is that, in the majority of cases, the method would lead to *some* conclusion that was true, and that in the individual case in hand, if there is any error in the conclusion, that error will get corrected by simply persisting in the employment of the same method. The validity of an inductive argument consists, then, in the fact that it pursues a method which, if duly persisted in, must, in the very nature of things, lead to a result indefinitely approximating to the truth in the long run. (CP, 2.781 [1902])

The primacy of quantitative induction lies in its ability to monitor the functioning of induction in general:

> The true guarantee of the validity of induction is that it is a method of reaching conclusions which, if it be persisted in long enough, will assuredly correct any error concerning future experience into which it may temporarily lead us. This it will do

> not by virtue of any deductive necessity (since it never uses all the facts of experience, even of the past), but because it is manifestly adequate . . . to discovering any *regularity* there may be among experiences, while *utter irregularity is not surpassed in regularity by any other relation of parts to whole*, and is thus readily discovered by induction to exist where it does exist, and the amount of departure therefrom to be mathematically determinable from observation where it is imperfect. (CP, 2.769 [c. 1905])

The use of statistical data to compile success ratios in order to assess the acceptability of theories provides the critical link between the qualitative and quantitative modes of induction. It becomes possible to use specifically quantitative induction to monitor the adequacy of inductive procedures as a whole, since every inductive generalization is correlated with the quantitative second-order thesis that it successfully accommodates in 100 percent of its applicative instances.

For Peirce, the only true test of the correctness of a theory is whether the inferences, applications, and predictions based on it prove successful. Purely intellectual factors such as explanatory power, parsimony, intuitive appeal, antecedent probability resulting from concordance with previously accepted theories, etc., are, for him, considerations relevant to the abductive process of selecting theories and hypotheses for testing. But these purely intellectual factors have no place in the specifically retroductive process of verifying the truth of theories and assessing their acceptability.

For Peirce, successful utilization affords the only strictly objective and nonintellectual standard for evaluating theories. In his Harvard lectures on pragmatism, he is explicit on its crucial controlling role:

> The justification for believing that an experiential theory which has been subjected to a number of experimental tests will be in the near future sustained about as well by further such tests as it has hitherto been, is that by steadily pursuing that method we must in the long run find out how the matter really stands. The reason that we must do so is that our theory,

> if it be admissible even as a theory, simply consists in supposing that such experiments will in the long run have results of a certain character. (CP, 5.170 [c. 1903])

It is clear how, for Peirce, quantitative induction validates induction in general—qualitative theory-projection included. Even a theory that has been completely successful in all tests to date may still prove inadequate in the long run, but further testing—that is, further quantitative inductions on as yet untested applications—will expose any error in our assessment and allow it to be corrected. We need assume only that the series of experiments by which we test a theory will reflect the nature of "reality" in the long run—an assumption which, for Peirce, is true by the very definition of the term, "reality."

Peirce's general position is exactly that of Hans Reichenbach, who writes:

> We thus come to the result that the rule of induction can by no means be maintained to be the best method of approximation. But with its help we can find better methods of approximation. Scientific method makes use of this fact to a great extent. . . . The rule of induction, or one of its equivalents, is the only method that can be used in the test of other methods of approximation, because it is the only method *of which we know* that it represents a method of approximation.[14]

Peirce, like Reichenbach, argues that if the experiential sector of nature has a "character," that is, if there is a reality in Peirce's sense (and this is something that cannot ever be fully established), then naive (straight rule) induction is bound to bring it out in the indefinitely prolonged long run.[15]

From this perspective, it is clear that even qualitative induction can be corrected through a quantitative monitoring of its products. Scientific progress is preeminently the change of adopted theories, and statistical controls can be used both (1) to determine that an improvement is necessary at this level (because the old theories are no longer 100 percent effective), and (2) to determine whether a proposed or supposed improvement is a real improvement (actually or probably). Accordingly, the process of qualitative induction itself can be correctively monitored by quantitative induction.[16] It can be subjected to quality control by statistical means. Peirce is thus at one with Sir Ronald Fischer in declaring that the theory of statistical design, and statistical inference in general, make key contributions to the theory of scientific induction.[17]

4. What Does Self-correctiveness Involve? Improving Our Theories versus Monitoring Them

Just how can an essentially statistical approach to the quality control of theorizing correct erroneous science so as to render the scientific enterprise self-corrective?

To begin with, it is necessary to note something it does not do. If we are incompetent in our theorizing, if the theories we accept are relatively inadequate, statistical checks will doubtless tell us that something is amiss, but they will certainly not do us the service of providing a better theory or theories to be adopted instead. Quantitative induction reveals *that* an improvement is necessary, but clearly not *which* improvement is needed.

The notion that science is self-corrective can be traced to several eighteenth-century writers (especially David Hartley [1705–1757], Georges Le Sage [1724–1803], and Joseph Priestley [1733–1804]), who took as their model the various mathematical methods of successive approximation, exemplified by such procedures as the well-known processes for determining *n*-th roots.[18] These methods, the rule of false position for example, operate in such a way that, given an initial position based on guesswork (however wild), there is an automatic procedure for successively revising this wrong answer into one that approximates stage by stage more closely to the correct one. As Priestley put it:

> Hypotheses, while they are considered merely as such, lead persons to try a variety of experiments, in order to ascertain

> them. These new facts serve to correct the hypothesis which gave occasion to them. The theory, thus corrected, serves to discover more new facts, which, as before, bring the theory still nearer to the truth. In this progressive state, or method of approximation, things continue. . . .[19]

Le Sage drew an analogy between the scientist at work and an arithmetician solving a problem in long division, producing at each stage a quotient more accurate than the one before.[20] Implicit throughout this eighteenth-century view is the conception that science possesses an *automatic and mechanically routine method* for continually improving on its older incorrect theories. This thesis—that science proceeds by routine steps of successive approximation inexorably closer to the truth, that scientific progress is a matter of convergence upon ''the correct answer''—is the initial form of the theory of self-correction.

This early view of the self-correctiveness of science in terms of an automatic algorithm for improving its claims once their deficiency becomes apparent, is of course untenable. When the progress of science indicates that an accepted hypothesis is not warranted, then neither ''the scientific method'' nor any other cognitive device affords any automatic way of producing a new and more appropriate replacement.

It cannot be said too emphatically that Peirce does not hold the ill-advised view that the method of science is ''self-corrective'' in providing some sort of automatic, cookbook procedure for devising good theories to put in place of bad ones.[21] Nothing in his teaching requires him to take this very dubious stance. The complaint that the statistical process of quantitative induction fails to establish the ''self-correctiveness'' of science in the specifically abductive sense of theory improvement (finding a new and better theory to put in place of one that has been impugned by statistical quality controls) is simply beside the point and does nothing to invalidate Peirce's theory of self-correctiveness.[22] To maintain that the methods of scientific inquiry will—or can—in the long run determine the truth of proposed answers to scientific questions, is not to say that science has a routine method (or anything approaching an automatically effective procedure) that enables it to ferret out the correct answers to its questions. Analogously, when it is said of the calculating prodigy that his genius enables him with a routine *procedure* (codifiable and transmittable to others) for solving such problems. Or again, when someone maintains that the free market provides a mechanism for pricing commodities, he does not thereby imply that it provides some explicit routine procedure for determining the market price of commodities.

Peirce's theory of self-correctiveness is simply that the methods of science in general provide the tools of theory-improvement, not that quantitative induction in particular does. Induction as a whole—the scientific method in general and not each of the several parts of induction—is "self-corrective," in the sense of self-monitoring.[23] The job of producing new theories is done by the scientific method at large, and not specifically and particularly by quantitative induction. Self-correctiveness in the sense of *performance-monitoring* does not proceed by induction as a whole, but by a part of induction (quantitative induction) for induction as a whole. Theory improvement is accomplished not by quantitative induction (or by any one part of scientific method separately), but by scientific (or inductive) methodology as a whole.

The crucial point is that the "self-correctiveness" of scientific inquiry has two aspects: (1) quality control or performance-monitoring, and (2) theory improvement. Quantitative induction takes care of the former and does so routinely in a mechanical and automatic way; the scientific method in general (induction as a whole) takes care of the latter, albeit in a way that is certainly not routine. With these distinctions understood, the resulting Peircean position is safe against the sorts of changes launched by the host of critics mentioned above.

5. Self-correctiveness Reexamined

But is there not still a serious problem here? Surely no Cartesian deity is at work issuing *a priori* guarantees that science will ultimately ferret out the real truth of a matter. Peirce's own beliefs to the contrary notwithstanding, science surely may, even in the eventual long run, arrive at a false position.[24] How then can it possibly be claimed to be self-corrective? The really crucial core of the claim that science is self-corrective is the profound Hegelian thesis that it is certainly not corrigible by anything else.

Science is autonomous. Corrections to science must come from science. Shortcomings in scientific work can be discovered only by further scientific work. The mistaken results of science can be improved or corrected only by further results of science. There can be no recourse at this point to tealeaf reading, numerology, the Delphic oracle, or the like. The self-correctiveness of science is best understood to lie in the final analysis simply in the autonomy of science, in its not falling subject to any external standard of correctness. Scientific claims must, whenever corrected at all, be corrected by further scientific claims. A ''science'' subject to external standards of correctness simply does not deserve the name of science: the truthfulness of scientific claims must be settled wholly within the scientific enterprise itself. This fundamental fact is the rock bottom on which the doctrine of the self-correctiveness of science must find its foothold.

The Peircean theory of science may be controversial in many respects, but the core of its doctrine that science is self-corrective (that is, autonomous and not admitting of any external correction) is a view that is inexorably pressed upon us by the groundrules of cognitive rationality, and one that is surely right.

6. Conclusion

Peirce's theory of the self-correctiveness of science is predicated on the far-reaching idea that the adequacy of our theorizing can be monitored by a key component of the scientific method itself; namely, by statistical controls applied to the results by putting our theories to work in various ways, especially in spectatorial (that is, predictive) and manipulative control over nature. By these standards our current theories are still far from perfect and science yet has a long way to go. But the superiority of, for example, the Pasteurian germ theory of disease over its earlier Galenic humor-imbalance counterpart does not rest on the evidence of its inner harmony, or intuitive appeal, systematic simplicity, or any other strictly intellectual factor. It has carried the day because of its statistical record of superior results in prediction and control over nature—in short, because of its greater applicative success. This idea—of a statistical monitoring of the success or failure of a theory's applications—is at the core of Peirce's conception of the self-corrective performance monitoring (that is, self-monitoring) of science. It provides an implicit and immanent standard of scientific progress, and a very reasonable one at that.[25]

2 Peirce on Scientific Progress and the Completability of Science

1. Peirce's Scientific Realism

Peirce propounded an ingenious theory regarding the relationship between the results of scientific inquiry and the nature of "the real truth," in factual matters that deal with actual existence in the world. In the face of the philosophic sceptic's agnosticism as to the very possibility of attaining "the real truth" about nature, Peirce proposed that *the truth is simply "the limit of inquiry," that is, what the scientific enterprise will discover in the idealized long run*, or would discover if the efforts were so extended.[26] Once scientific progress reaches a point at which a question is answered in a certain way and that answer is thereafter maintained without change within the ongoing community of inquirers, then it is indeed the true answer to the question in hand.[27] At issue is a Copernican inversion reminiscent of Kant: It is not that "rational inquiry" is appropriate because what it ultimately arrives at is the actual truth, but that "the actual truth" qualifies as such because rational inquiry ultimately arrives at it. As Peirce put it, the truth simply *is* "the opinion which is fated to be ultimately agreed to by all who investigate."[28] Accordingly, he maintained:

> Reality is that mode of being by virtue of which the real thing is as it is, irrespectively of what any mind or any definite collection of minds may represent it to be. The truth of the

> proposition that Caesar crossed the Rubicon consists in the fact that the further we push our archaeological and other studies, the more strongly will that conclusion force itself on our minds forever—or would do so, if study were to go on forever. (CP, 5.565 [1901])

Truth is *adaequatio ad rem,* and reality answers to the view of things lying at the *focus imaginarius* of the end of inquiry. As Peirce put it:

> [R]eality, the fact that there is such a thing as a true answer to a question, consists in this; that human inquiries,—human reasoning and observation,—tend toward the settlement of disputes and ultimate agreement in definite conclusions which are independent of the particular stand-points from which the different inquirers may have set out; so that the real is that which any man would believe in, and be ready to act upon, if his investigations were to be pushed sufficiently far. (CP, 8.41 [c. 1885])

By this ingenious doctrine of Hegelian inspiration, Peirce was able to bridge the Kantian gap between reality as it is *an sich* and *our reality* (our scientific conception of reality) by holding that the former is simply the long-range projection of the latter.[29]

Specifically, Peirce's theory was predicated on two contentions:

1. *Ultimate Correctness*. Whatever science will come to maintain over the theoretical long run is indeed true. (Over the TLR, science maintains *only* the truth.)
2. *Ultimate Completeness*. *All* truth regarding the world will be realized by science in the theoretical long run. (Over the TLR, science maintains *all* the truth.)

The first of these theses stipulates the accuracy of theoretical-long-run science, its ability to get at the actual truth of things.[30] The second stipulates its comprehensiveness (nay, potential omniscience), that no general truths about the world are in principle beyond its ken, that nature contains no ultimately occult compartments of inaccessible truth.[31] Like a good witness, theoretical-long-run science presents the truth, the whole truth, and nothing but the truth.

On such a view, the truth about factual matters at the level of generality simply *coincides* with what science will maintain to be so in the long run. Genuine factual knowledge, our attainment of "the real truth" about the world, can thus be construed as being what Peirce characterized as the "final irreversible opinion" of the scientific community.[32] Espousal in the long run within an ideally projected community of scientific inquirers is both a necessary and sufficient condition for the truthfulness of generalizations in the factual (world-oriented) domain. For Peirce, the decisive consideration in the determination of truth is stability; Peirce himself used the word "fixity." According to him, the mark of a true belief, one constrained by an external and independent reality, is that it is destined or, as he put it, "fated" to be underwritten by the operation of scientific method.[33]

2. The Historical Background of Peirce's Doctrine: The *Fin de Siècle* View of Natural Science

To clarify Peirce's position, let us examine it in its historical context.

One acute contemporary analyst of physics speculates as follows about the ultimate completion of his field:

> It is possible to think of fundamental physics as eventually becoming complete. There is only one universe to investigate, and physics, unlike mathematics, cannot be indefinitely spun out purely by inventions of the mind. The logical relation of physics to chemistry and the other sciences it underlies is such that physics should be the first chapter to be completed. No one can say exactly what completed should mean in that context, which may be sufficient evidence that the end is at least not imminent. But some sequence such as the following might be vaguely imagined: The nature of the elementary particles becomes knwon in self-evident totality, turning out by its very structure to preclude the existence of hidden features. Meanwhile, gravitation becomes well understood and its relation to

> the stronger forces elucidated. No mysteries remain in the hierarchy of forces, which stands revealed as the different aspects of one logically consistent pattern. In that imagined ideal state of knowledge, no conceivable experiment could give a surprising result. At least no experiment could that tested only fundamental physical laws. Some unsolved problems might remain in the domain earlier characterized as organized complexity, but these would become the responsibility of the biophysicist or the astrophysicist. Basic physics would be complete; not only that, it would be manifestly complete, rather like the present state of Euclidean geometry.[34]

Extended from physics to natural science in general, such a position holds that the realm of potential discovery is effectively exhaustible. To be sure, it is never wholly and completely exhausted: the total exploration of nature is achieved only gradually and "in the limit," because the upper limit of potential scientific discovery is never actually reached but only approached asymptotically over the long run of scientific progress. Nevertheless, this perspective encourages a situation in which really big surprises remain. To be sure, there is no final *Gotterdämmerung*, and natural science does not come to a stop, but really important innovation fades away for all practical purposes.

A position of just this sort was maintained by Peirce who, in effect, saw the history of science as progressing by way of asymptotic approximation to a finally adequate picture of the world. He wrote that we can expect "in the progress of science its error will indefinitely diminish, just as the error of 3.14159, the value given for π, will indefinitely diminish as the calculation is carried to more and more places of decimals."[35]

As Peirce saw it, the current stage of scientific knowledge is such that further scientific progress is solely a matter of increasing accuracy, of filling in with increasing refinements and ever more accurate determination the details of a picture whose shape becomes increasingly clear and well-defined. Future progress is a matter of providing increasingly fine-grained detail within a context whose course-grained structure has already been determined. Thus in about 1896,

Peirce wrote, that it is not implausible to think, given the content of current science, that "the universe is now entirely explained in all its leading features; and that it is only here and there that the fabric of scientific knowledge betrays any rents."[36]

Peirce was by no means alone in regarding such a view as tenable. In physics above all, the conviction was widespread in the 1875–1905 era that the days of major innovations were over, that all the really big discoveries had been made. Some of the most able physicists of the day shared this sense, that the discipline had reached its more or less completed form and that little remained to be done, apart from work on relatively minor issues.

In a dedication address delivered at the Ryerson Physical Laboratory at the University of Chicago in 1894, A. A. Michelson, America's first Nobel Laureate in science, remarked:

> While it is never safe to affirm that the future of Physical Science has no marvels in store even more astonishing than those of the past, it seems probable that most of the grand underlying principles have been firmly established and that further advances are to be sought chiefly in the rigorous application of these principles to all the phenomena which come under our notice.
>
> It is here that the science of measurement shows its importance—where quantitative results are more to be desired than qualitative work. An eminent physicist has remarked that the future truths of Physical Science are to be looked for in the sixth place of decimals.[37]

Michelson's gloomy forecast was echoed by T. C. Mendenhall, formerly a physics professor, at the time a college president, and soon to be both president of the American Association for the Advancement of Science and Superintendent of the United States Coast and Geodetic Survey. In his popular text on electricity (1887), he maintained:

> More than ever before in the history of science and invention, it is safe now to say what is possible and what is impossible. No one would claim for a moment that during the next five

> hundred years the accumulated stock of knowledge of geography will increase as it has during the last five hundred. . . . In the same way it may safely be affirmed that in electricity the past hundred years is not likely to be duplicated in the next, at least as to great, original, and far-reaching discoveries, or novel and almost revolutionary applications.[38]

This *fin de siècle* sentiment that the great heroic deeds of physical science were over was by no means confined to a few eccentric notables.[39] In his 1956 presidential address to the American Physical Association, R. T. Birge recalled his first physics teacher at the University of Wisconsin in 1906:

> To him physics was an incomparably beautiful, but *closed* subject. There was nothing in his lectures to suggest that there were things still to be discovered in physics, and hence no incentive to enter the field except to become a teacher and in turn show these same beautiful experiments to one's own students.[40]

An even more remarkable instance of the same phenomenon is given by Max Planck:

> As I was beginning to study physics [in 1875] and sought advice regarding the conditions and prospects of my studies from my eminent teacher Phillip von Jolly, he depicted physics as a highly developed and virtually full-grown science, which—since the discovery of the principle of the conservation of energy had in a certain sense put the keystone in place—would soon assume its finally stable form. Perhaps in this or that corner there would still be some minor detail to check out and coordinate, but the system as a whole stood relatively secure, and theoretical physics was markedly approaching that degree of completeness which geometry, for example, had already achieved for hundreds of years. Fifty years ago [as of 1924] this was the view of a physicist who stood at the pinnacle of the times.[41]

Contemplation of the enormous strides being made all across the scientific frontier—in biology, medicine, chemistry, and on and on—led to the widespread if not typical belief that science stood pretty much at the last frontiers and that the course of progress in scientific knowledge—so dramatically explosive since its first great flourishing in the seventeenth century—was not approaching completion. From the teacher of Planck in the 1870s to the teacher of Birge in the 1900s, a substantial group among those working physicists who thought about the issue at all took the view that the potential range of physical knowledge is finite and moreover held that the proportion of the known to the unknown sector of this finite range was relatively large.

In this regard, Peirce was a child of his time. But the predominant ethos was one of success and self-congratulation, decidedly not of failure. To this extent, to speak of *fin de siècle*, with its overtones of weariness, exhaustion and failure of nerve, is not appropriate. The dominant sentiment in metascientific theory was elation, even pride, in the immense strides taken, a sense of power approaching *hubris* that the intellectual conquest of nature was virtually complete.[42] As Peirce wrote in 1878: "the new phenomena which now remain to be discovered are probably only of secondary importance."[43]

3. The Foundation of Peirce's Theory: The Geographic Exploration Model

Peirce's equation of "the real truth" about nature with the ultimate conclusions of science in the idealized long run immediately encounters an apparent difficulty. It seems plausible to suppose *a priori* that the physical world does have a nature of some sort, that something or other is bound to be generally true of it. And if we assume, with Peirce, that all truth is (ultimately) discovered truth stably maintained, then problems arise. For how can we possibly know in advance that science finally will settle down to a stable teaching? And so various critics take umbrage when Peirce insists that "there is a general *drift* in the history of human thought which will lead it to one general agreement, one catholic consent."[44] Let Bertrand Russell speak for them all:

> Is this an empirical generalization from the history of research? Or is it an optimistic belief in the perfectability of man? Does it contain an element of prophecy or is it a merely hypothetical statement of what would happen if men of science grew continually cleverer? Whatever interpretation we adopt, we seem committed to some very rash assertion.[45]

What guarantee have we that scientific progress will ultimately lead to a stable result, that scientific opinion will not eventually simply diverge or oscillate or go off on a long random walk? Peirce thinks that we do have such an assurance, and his position on this question is indeed crucial to his metaphysics of knowledge.

Briefly stated, Peirce's doctrine amounts to a *cumulative-convergence theory* of scientific progress.

Peirce, in effect, saw the history of science as progressing through two stages: an initial or preliminary (noncumulative) phase of groping for the general structure of the *qualitative* relations among the parameters of nature, and a secondary (cumulative) phase of *quantitative* refinement, where the second phase would determine with increasing precision the exact values of parameters that figured in equations whose general configuration was determined in the initial one. In the first stage, there is the qualitative change of a succession of substantively discordant theories. In the second stage, the main qualitative issues (we might say, the equation-forms) are settled, and it becomes a matter of refining knowledge of the constants at issue. Once the first qualitative phase is completed—a stage Peirce believed to have been realized by his own day, at least in the physical sciences[46]—then ongoing scientific progress is just a matter of increasing detail and exactness, of determining the ever more minute decimal-place values of quantities whose approximate value is already well established. The increasingly sophisticated radiotransmission of photographs provides an apt analogy. There is indeed change, but such change always preserves the rough outline of the old while filling in its details.[47]

On this view, science in due course enters into a condition where further progress is cumulative (always preserving and merely improving on what has gone before) and convergent (further improvements come to be smaller and smaller in scope).[48] Indeed, convergence presupposes a cumulation of sorts. For if there is to be convergence, then the previous position must always be preserved in its essentials and then improved in some relative detail. Convergence implies reaching a stage after which further change becomes a matter of filling in yet another decimal place to lend added refinement to an already largely fixed picture. Accordingly, Peirce insists:

> [S]cience does not advance by revolutions, warfare, and cataclysms, but by coöperation, by each researcher's taking advantage of his predecessors' achievements, and by his joining his own work in one continuous piece to that already done. (CP, 2.157 [1902])

On such a view, science undeniably "has a future" since there will always be worthwhile discoveries to be made. But discoveries take place within a context of overall limits, and the magnitude of their intrinsic importance becomes ever smaller. New discoveries will not, nay cannot, modify the overall understanding of processes of nature in any fundamental way. They serve to increase the accuracy or sophistication of a basically determinate view of the world, making adjustments and refinements which, however difficult and important the work of discovery itself might be, still produce only marginal adjustments in our intellectual world picture. To be sure, scientific progress will not ever quite reach a completed, final, static, and unchanging state, because the realm of potential discovery is never completely mapped. Rather, it moves toward this position by way of asymptotic approximation to a finally adequate picture of the world.

The analogy of geographic exploration is suggestive here. First an entire hemisphere is added to the globe. Then oceans and continents are explored. After that, the source of the Nile and the depths of the Arabian desert are brought to light. Eventually the North Pole is visited and Everest is scaled. The latter are great, perhaps monumental, *achievements*, but as *discoveries* their magnitude shrinks to comparative insignificance.[49]

This geographic-exploration model is not a historical curiosity that flourished in Peirce's day and died thereafter. It continues to thrive today. Recently, it was given a clear and eloquent formulation by the microbiologist, Gunther S. Stent:

> I think everyone will readily agree that there are *some* scientific disciplines which, by reason of the phenomena to which they purport to address themselves, are *bounded*. Geography, for instance, is bounded because its goal of describing the features of the Earth is clearly limited. . . . And, as I hope to have shown in the preceding chapters, genetics is not only bounded, but its goal of understanding the mechanism of transmission of hereidtary information *has*, in fact, been all but reached. Indeed, and here I will probably part company with some who might have granted me the preceding example, even such much more broadly conceived scientific taxa as chemistry and biology are also bounded. For in the last analysis, there is immanent in their aim to understand the behavior of molecules and of "living" molecular aggregates a definite, circumscribed goal. Thus, though the total number of possible chemical molecules is very great and the variety of reactions they can undergo vast, the goal of chemistry of understanding the principles governing the behavior of such molecules is, like the goal of geography, clearly limited. . . . [T]here is immanent in the evolution of a bounded scientific discipline a point of diminishing returns; after the great insights have been made and brought the discipline close to its goal, further efforts are necessarily of ever-decreasing significance.[50]

We have here an accretional view of the progress of science, with each successive accretion inevitably making a relatively smaller contribution to what has come before. Progress, on this view, consists in driving questions down to lesser and lesser magnitudes, providing increasingly enhanced detail of increasingly diminished significance.[51] This at bottom is the Peircean vision of ultimate convergence in scientific inquiry.[52]

4. A Critique of the Geographic Exploration Model

The plausible analogy of geographic exploration is nevertheless fundamentally mistaken. It views scientific progress as a whole on the basis of one particular (and by no means typical) sort of progress, the sequential filling in of an established framework with greater and greater detail, lending additional refinement to a fundamentally fixed result. This view combines two gravely erroneous ideas: (1) that science progresses by cumulative accretion (like the growth of a coral reef), and (2) that the magnitude of these additions is steadily decreasing.

If the first of these ideas collapses, so does the position as a whole. And collapse it does, for science progresses not just *additively* but in large measure also *subtractively*. As Thomas Kuhn and others have persuasively argued, today's most significant discoveries always represent an overthrow of yesterday's: the big findings of science inevitably take a form that contradicts its earlier big findings and involve not just supplementation but replacement. Substantial headway is made preeminently through conceptual and theoretical innovation. The preservationist stance, that the old views were accpetable as far as they went and merely need supplementation, will not serve. Significant scientific progress is genuinely revolutionary in that there is a fundamental change of mind about how things happen in the world.

The medicine of Pasteur and Lister does not add to that of Galen or of Paracelsus, but replaces them. The creative scientist is as much a demolition expert as a master builder. Significant scientific progress is generally a matter not of adding further facts on the order of filling in of a crossword puzzle but of changing the framework itself. Science in the main develops not by addition but by substitution and replacement.[53] Progress lies not in a monotonic accretion of more information but in superior performance in prediction and control over nature.[54]

The Peircean doctrine of convergent cumulation must, accordingly, be abandoned. Nevertheless, it must be recognized that Peirce contributed an insight of immense value in this sphere. This insight concerns the economic aspect of the matter. Let us look at it more closely.

5. Cost-Escalation versus Yield-Diminution

In his pioneering 1879 essay on "Economy of Research," Peirce addressed the issue of the increasing difficulty of scientific progress:

> We thus see that when an investigation is commenced, after the initial expenses are once paid, at little cost we improve our knowledge, and improvement then is especially valuable; but as the investigation goes on, additions to our knowledge cost more and more, and, at the same time, are of less and less worth. Thus, when chemistry sprang into being, Dr. Wollaston, with a few test tubes and phials on a tea-tray, was able to make new discoveries of the greatest moment. In our day, a thousand chemists, with the most elaborate appliances, are not able to reach results which are comparable in interest with those early ones. All the sciences exhibit the same phenomenon, and so does the course of life. At first we learn very easily, and the interest of experience is very great; but it becomes harder and harder, and less and less worthwhile. . . . (CP, 7.144)

As this passage makes clear, Peirce sees natural science as subject to (1) rising costs and (2) diminishing returns.

It is important to realize that two substantially different ideas are conjoined in this Peircean picture of the economics of scientific progress. The first of these is the ideal of *cost-escalation*: the conception that with the progress of science, it becomes increasingly expensive (in both material resources and human effort) to achieve worthwhile new results. The second is *yield-diminution*: the conception that with the progress of science the later findings are inevitably of an increasingly diminished significance. Diagramatically, the contrast stands as follows:

1. later → harder (more difficult, more expensive)
2. later → lesser (less significant)

To make use of a homely apple-picking analogy, the first thesis says that the later apples are harder to get off the tree than the earlier; the second says that the later apples are smaller than the earlier.[55] These two ideas move in quite different directions. Though conjoined in Peirce's thought, they are certainly separable.

The point to be stressed in this connection is that the preceding critique of Peirce's position relates solely to the second member of this pair, his conception of yield-diminution. The strictures presented above to not touch the thesis of cost-escalation, an idea which seems to me profoundly insightful and entirely right. This important point warrants closer scrutiny.

6. Cost-Escalation

Evidence of the escalating costs of scientific progress is provided by some suggestive statistics:[56]

1. In the U.S.A. the total (real) expenditure per scientist has been increasing at a rate of roughly 7 percent per year (thus doubling roughly every ten years) throughout recent history,[57] while the per-capita productivity of scientists (measured by contributions to the literature) has remained relatively constant.
2. The number of scientists working in the U.S.A. has been increasing at roughly 6 percent per year (doubling every twelve years or so), whereas (a) the number of "eminent" men (those selected for listing in the standard biographical handbooks or other registers that select only a limited elite of scientific contributors) has been increasing at a rate of only

> about 3 percent per year (thus doubling in around twenty years),[58] and (b) the number of relatively significant findings (those cited in the references of synoptic monographs, handbooks, and textbooks) has to all appearances been growing at a virtually *linear* rate.[59]

The historic situation regarding the costs of American science was carefully delineated by Raymond Ewell in 1954.[60] His study of research and development expenditures in the U.S. showed that growth has been exponential; from 1776 to 1954 nearly $40 billion was spent, and half of that was spent after 1948. Research and development expenses were found to be increasing at a rate of 10 percent per year.[61] Projected at this rate, Ewell saw the total climbing to what he viewed as an astronomical $6.5 billion by 1965, a figure that turned out to be too conservative.[62] By the mid-1960s, America was spending an amount on research and development that was more than the whole of the federal budget before Pearl Harbor.

The proliferation of scientific facilities has proceeded at an impressive pace over the past hundred years. In the early 1870s there were only eleven physics laboratories in the British Isles; by the mid-1930s there were more than three hundred;[63] today there are several thousand. The scale of activities in these laboratories has also expanded vastly. It is perhaps unnecessary to dwell on the immense cost of the research equipment of contemporary science. Even large organizations can hardly keep pace with rising levels of research expenditures.[64] Radiotelescopic observatories, low-temperature physics, research hospitals, and lunar geology all involve outlays on a scale that require the funding support of national governments, sometimes even consortia of governments. Science has increasingly become a very expensive undertaking. Alvin M. Weinberg, former Director of the Oak Ridge National Laboratory, has written:

> When history looks at the 20th century, she will see science and technology as its theme; she will find in the monuments of Big Science—the huge rockets, the high-energy accelerators, the high-flux research reactors—symbols of our time just as surely as she finds in Notre Dame a symbol of the Middle Ages.[65]

Every recent statistical study that has been made of the costs of scientific research projects in private industry, in government, and in academic institutions yields the uniform result that the per project cost (in real dollars) has grown at a doubling time of less than ten years throughout recent decades. Unless we are prepared to accept the farfetched view that the average unit yield of scientific research has been improving impressively, we are driven to a picture of steadily increasing human and material costs of high-level scientific results. There are, it seems, substantial grounds for agreement with Max Planck's appraisal:

> To be sure, *with every advance* [in science] *the difficulty of the task is increased; ever larger demands are made on the achievements of researchers,* and the need for a suitable division of labor becomes constantly more pressing.[66]

It emerges that the economic sector of Peirce's theory of scientific progress involves two, distinct, though for him interrelated, contentions, one of which, yield-diminution, deserves to be scrapped, but the other of which, cost-escalation, represents an important insight into the structure of scientific research.

7. An Economic Critique of Peirce's Theory of the Truth-Science Relationship

It is somewhat ironic that Peirce's scientific realism can be seen to be flawed from a perspective that is in fact highly congenial to it—an economic one.

A zero-growth world is upon us in science as elsewhere. The resources at our disposal are limited, and we shall not be able to continue to exploit them at exponentially increasing levels as we have done in the past. There is, for example, a limit—a fundamentally economic limit—to the size of the particle accelerators, radio telescopes, high-flux reactors, etc., that can be constructed. These limits inexorably circumscribe our cognitive access to the real world. There are interactions with nature of such a scale (measured in such parameters as energy, pressure, temperature, particle velocities, etc.)

whose realization would require the use of resources on so vast a scale that we could never realize them.

But if there are interactions to which we have no access, then there are (presumably) phenomena that we cannot discern. It would be unreasonable to expect that nature has confined the distribution of phenomena of potential cognitive significance to ranges that lie within our ken.

Where there are inaccessible phenomena, there must be cognitive incompleteness. To this extent, at any rate, the empiricists were surely right. Moreover, if certain phenomena are not just undetected but by their very nature are inaccessible to us (even if only for the merely economic reasons noted above), then our theoretical knowledge of nature must presumably be incompletable. Only the most hidebound rationalists could uphold the capacity of sheer intellect to compensate for the lack of data. Where there are unobserved phenomena, we must reckon with the prospect that our theoretical systematizations are incomplete.

Thus, there is a limit, ultimately an economic limit, to the questions about the phenomena of nature that we can ever answer. Many questions of transcending scientific importance will remain unresolved because their resolution would demand a greater concurrent deployment of resources than will ever be marshalled at any one time in a zero-growth world. There will thus be truths that science cannot attain, so that we shall not (even in the infinite long run) achieve "the whole truth." The economic requisites of scientific work are such that our knowledge of nature must finally remain incomplete.

Think once again of the main contentions of the preceding discussion: (1) in science, yield-diminution must be rejected because major new scientific innovations—revolutionary discoveries—are always possible in principle, and (2) that cost-escalation must be accepted as a fact of scientific life; scientific revolutions, though possible, are increasingly more difficult and expensive to mount. These contentions have dire repercussions for a Peircean scientific realism. Consider again his two key theses:

1. *Correctness*: That whenever science will come to maintain

over the theoretical long run (TLR) is true; over the TLR, science maintains nothing but the truth.

2. *Completeness*: That all truth regarding the world will be realized by science in the theoretical long run; over the TLR, science maintains all the truth.

Our present consideration of the economics of scientific progress cast a strong shadow of dubiousness over Peirce's position. His completeness thesis emerges as untenable because, for essentially economic reasons, there will be questions that science will never be able to resolve, even in the theoretical long run.[67] It would thus appear that certain features fundamental to the very structure of man's inquiry into the ways of the world conspire to limit the knowledge that we can attain in this sphere.[68]

Even Peirce's correctness thesis must be rejected, because what science maintains over the indefinitely projected long run could well be false, having defects that could be discovered only on the far side of an economic data-barrier. The effort needed to determine the falsity of claims at issue might make demands on resources that could not be met in a zero-growth world. There is every reason to deny that what we ultimately reach is actually "nothing but the truth" since there is every reason to think that where scientific knowledge is concerned further knowledge does not just supplement but generally corrects our knowledge-in-hand, so that the incompleteness of our information entails its incorrectness as well.

8. The Economic Impediments to Peirce's Scientific Realism

Peirce overlooked a crucial factor when he wrote:

> [T]hought, controlled by a rational experimental logic, tends to the fixation of certain opinions, equally destined, the nature of which will be the same in the end, however the perversity of thought of whole generations may cause the postponement of the ultimate fixation. (CP, 5.430 [1905])

Whatever consensus may arise from "rational experimental logic," in natural science we must work with data. These data stem from interactions with nature, and the realization and cognitive exploitation of these interactions has an economic aspect that limits their completeness. As anyone who has ever confronted a curve-fitting problem well knows (Peirce himself certainly included), incomplete data lend themselves to incompatible extrapolations. In the very nature of the case, an ultimate fixation of opinion is unattainable in such circumstances.

Peirce's theory of the science-truth relationship is defective simply because it ventures too far into abstraction from the human limitations within which scientific work must actually be done. His conception may well be plausible—for a community of powerful disembodied intelligences whose capabilities and efforts at mastering the resistances of nature are unimpeded by any economic constraints, beings whose observations are obtained by cost-free processes, for whom computations and data-processing are to be had for the asking, or whose experimental interactions with nature can be carried out by acts of will alone. But creatures like us, whose scientific efforts are subject to crucial limitations of resources, cannot be confident that their science must in principle ultimately ferret out "the real truth." In the actual circumstances, the Peircean vision of the scientific enterprise is visionary, in the sense of being unrealistic.

Peirce and those scientific realists who follow him hold the ill-advised view that only the limitations of time separate "the-truth-about-reality" from "the teachings of science." They see truth attainment as a one-factor idealization looking to the removal of a single limitation—that of effort maintained over a sufficiently extended course of time. However, the present deliberations indicate that if one wants to move along such a route, then one must resort at least to a two-factor idealization with respect not only to time but (more drastically) also to the availability of resources. And the presence of this second dimension of idealization introduces an unrealistic aspect of far-fetchedness into the scientific realism at issue, rendering the whole thing a very dubious proposition. For while the prospect of intelligent life on other planets may conceivably remove temporal limitations (cf. footnote 27 above), it cannot remove economic ones.

These considerations indicate that any version of a scientific realism of Peircean stamp is untenable when it proposes to conceptualize reality as "what will eventually be held to be the case by science (over the long run)." If there is reason to think—in line with our theory—that even the ultimate scientific consensus (if there was to be one) would inevitably be an imperfect fragment of some larger, more complex structure that lies further down the road than we can ever actually afford to travel, then the equation:

reality as it actually is = reality as science will ultimately (and irreversibly) deem it to be

must be abandoned. One cannot venture to close the gap between our putative reality and genuine reality by resort to the contrast between what we think at some juncture and what we are fated to think in the long-run (as Peirce puts it).

To be sure, this critique does not require us to abandon Peirce's conception of the essentially experiential character of "reality." But, as he himself insists, it is not actual but possible experiences that are at issue, not just the *is*s but also the *would-be*s of the thing. And economic limitations of the sort we have noted are crucial here, placing an entire range of experiential *would-be*s outside our effective reach, not just until further notice, but permanently. A sector of (theoretically) experientiable reality becomes *de facto* inaccessible.

On such a view, the progress of science, even at the level of the idealized long run, will never issue in authentic reality, but only in a historical succession of suboptimally putative realities, reality-pictures to which we cannot impute any sort of finality, not even that merely approximate finality envisaged by Peirce.

Accordingly, we seem driven to the conclusion that there is no real justification for equating "the truth about reality" with "the teachings of science," if the science at issue is indeed our human science, carried on by creatures subject to the fundamentally inevitable economic limitation of finite resources.

This upshot is ironic, as we have noted, because Peirce was in many contexts so acutely (and pioneeringly) alive to the economic aspects of science. It was Peirce himself who brought the issue of the economics of science to the forefront and virtually single-handedly laid the foundations for this enterprise in the discipline he characterized as "the economy of research."[69] His basic insights into the economic aspects of science, particularly his views on the cost-escalation of ongoing significant scientific work, are prescient, valid, and highly useful for a proper understanding of the scientific enterprise. Yet these very considerations operate to cast strong doubt upon the tenability of Peirce's metaphysical teaching that science is fated to arrive at the real truth of things in the long run.

The line of objections considered here did not wholly escape Peirce himself, and he came close to dealing with it. In the important unpublished 1884 paper, "Design and Chance" (Ms. 875), Peirce proceeds to "call in question," and indeed to "call into doubt," his own earlier doctrine—the "fundamental axiom of logic" that "every intelligible question whatever is susceptible in its own nature of receiving a definitive and satisfactory answer, if it be sufficiently investigated by observation and reasoning" (p. 7).[70] And in one extremely interesting passage, typical of his thinking on this issue after 1880, he maintained:

> [I]f we think that some questions are never going to get settled, we ought to admit that our conception of nature as absolutely real is only partially correct. Still, we shall have to be governed by it practically; because there is nothing to distinguish the unanswerable questions from the answerable ones. . . . (CP, 8.43 [c. 1885].)[71]

In the context of this passage, Peirce contemplates a retreat from his usual hopeful view that whatever can ultimately be answered by science will ultimately be answered, that "there is an ascertainable true answer to every intelligible question."[72] This withdrawal leads him to the uncomfortable and un-Peircean thesis that perhaps "our conception of nature as absolutely real is only partially correct." To be sure, Peirce continues to hold tight to the central doctrine of his metaphysics that all of reality will ultimately be known,[73] but now the prospect of ultimately unanswered questions about nature opens a disconcerting gap between (natural) reality on one hand and nature itself on the other,[74] a gap between empirical and noumenal "reality" that reopens the Kantian issue that Peirce's theory of truth was designed to close. It is doubtful in the extreme whether we—or indeed Peirce himself, struggle though he might[75]—could long manage to live with such a gap.

We are happily under no compulsion to occupy this uncomfortable position. After all, to vindicate the methods of science as tools of inquiry we do not need to invoke the cumulative-convergence doctrine, and we do not need an *a priori* guarantee—not even a regulative assumption—that science will ultimately "deliver the goods" regarding the real truth of things. All that we do need is a reasonable assurance that by adopting the methods of scientific inquiry we shall do as well as it is possible to do in the epistemic circumstances of the case, that the methodological posture of science is in no way inferior to its contemplatable alternatives. And this sort of defense is in fact one that the whole gamut of Peircean considerations—the appeal to autonomous self-correctiveness, metaphysical soundness, and applicative success—is well suited to provide.[76]

3 Peirce on Abduction, Plausibility, and the Efficiency of Scientific Inquiry

PART ONE: PLAUSIBILITY AND ABDUCTIVE TALENT

1. The Centrality of Abduction

The taxonomic structure of methodology on Peirce's conception of science would look like this:

- inductive methodology of science
  - quantitative induction
  - qualitative induction
    - abduction (hypothesis formulation and selection)
    - retroduction (hypothesis testing and elimination)

Here quantitative induction is (in effect) statistics, and qualitative induction is a method of hypotheses. Peirce's understanding of the hypothetico-deductive model of scientific inquiry is thus based on the appropriate meshing of two processes: abduction (hypothesis projection) and retroduction (hypothesis elimination). By the former, inquiring scientists engage in the conjectural proliferation of explanatory hypotheses that merit closer scrutiny; by the latter they proceed to reduce them by empirical testing.[77] This overall view of the inductive process if virtually indiscernible from the conjecture-and-refutation model of scientific inquiry advocated by K. R. Popper in the present century.

The task of abduction is to determine a limited area of promising possibility within the overall domain of theoretically available hypotheses, a region which is at once small enough for detailed examination and research, and large enough to afford a good chance of containing the true answer. Abduction determines the research-worthy hypotheses; retroduction puts them through the eliminative screening of actual testing. Peirce described abduction as follows:

> *Presumption*, or, more precisely, *abduction* . . . furnishes the reasoner with the problematic theory which induction verifies. Upon finding himself confronted with a phenomenon unlike what he would have expected under the circumstances, he looks over its features and notices some remarkable character or relation among them, which he at once recognizes as being characteristic of some conception with which his mind is already stored, so that a theory is suggested which would *explain* (that is, render necessary) that which is surprising in the phenomena. He therefore accepts that theory so far as to give it a high place in the list of theories . . . which call for further examination. . . . Presumption is the only kind of reasoning which supplies new ideas, the only kind which is, in this sense, synthetic. (CP, 2.776–777 [1902])

As this discussion indicates, the abduction/retroduction model of inductive inquiry embodies a characteristic difficulty. How is it that the human mind possesses a manageable procedure for sampling which enables it to draw in short order relatively good hypotheses from an infinite population including mostly relatively bad ones? Conjectural fancy is limitless, but resources are scarce and life is short. Even though abduction is not a matter of accepting a hypothesis, but of its being "entertained interrogatively," as Peirce puts it,[78] possibilities cannot in practice be spun out forever. Given that the scientist cannot hope to scrutinize, let alone test, every imaginable hypothesis, how is he to proceed in the face of limitless possibilities, considering that he is a creature of finite capabilities and means? Can certain prospects validly be dismissed out of hand and others established as deserving further attention more or less *a priori*, without detailed analysis and evaluation?

Peirce thinks that this can indeed be done:

> We shall do better to abandon the whole attempt to learn the truth, however urgent may be our need of ascertaining it, unless we can trust to the human mind's having such a power of guessing right that before very many hypotheses shall have been tried, intelligent guessing may be expected to lead us to the one which will support all tests, leaving the vast majority of possible hypotheses unexamined. Of course, it will be understood that in the testing process itself there need be no such assumption of mysterious guessing-powers. It is only in selecting the hypothesis to be tested that we are to be guided by that assumption. (CP, 6.530 [c. 1901])

Peirce goes on to explain this "power of guessing right":

> [I]t is a primary hypothesis underlying all abduction that the human mind is akin to truth in the sense that in a finite number of guesses it will light upon the correct hypothesis. . . . For if there were no tendency of that kind, if when a surprising phenomenon presented itself in our laboratory, we had to make random shots at the determining conditions, trying such hypotheses as that the aspect of the planets had something to do with it, or what the dowager empress had been doing just five hours previously, if such hypotheses had as good a chance of being true as those which seem marked by good sense, then we never could have made any progress in science at all. But that we have made solid gains in knowledge is indisputable; and moreover, the history of science proves that when the phenomena were properly analyzed . . . it has seldom been necessary to try more than two or three hypotheses made by clear genius before the right one was found. (CP, 7.220 [c. 1901])

Here, as Peirce sees it, we come to the very crux of the matter: the "problem of *induction*" reduces to the "problem of *abduction*"—the question of how it is that the human mind is successful at hypothesis formulation. How is it that we can arrive at true (or, at any rate, promising, i.e., nearly true) hypotheses, given that there are always infinitely many grossly wrong ones from which to choose?

2. Peirce's Concept of Plausibility

According to Peirce the guidance needed for effective theorizing issues from considerations of plausibility. To implement this idea he advances a distinctive doctrine of the plausibility of theses and theories regarding how things work in the world.

As Peirce sees it, man is endowed through the evolutionary process not only with the instincts of an animal and with everyday "common sense" or "horse sense" but also with their functional equivalent in the cognitive domain, a sense of the plausible regarding the workings of nature:

> Every inquiry whatsoever takes its rise in the observation . . . of some surprising phenomenon, some experience which either disappoints an expectation, or breaks in upon some habit of expectation. . . . At length a conjecture arises that furnishes a possible Explanation. . . . On account of this Explanation, the inquirer is led to regard his conjecture, or hypothesis, with favor. As I phrase it, he provisionally holds it to be "Plausible"; this acceptance ranges in different cases—and reasonably so—from a mere expression of it in the interrogative mood, as a question meriting attention and reply, up through all appraisals of Plausibility, to uncontrollable inclination to believe. (CP, 6.469 [1908])

He goes on to explicate the nature of plausibility:

> By plausibility, I mean the degree to which a theory ought to recommend itself to our belief independently of any kind of

> evidence other than our instinct urging us to regard it favorably. All the other races of animals certainly have such instincts; why refuse them to mankind? . . . Physicists certainly today continue largely to be influenced by such plausibilities in selecting which of several hypotheses they will first put to the test. (CP, 8.223 [c. 1910])

This sense of the plausible endows the scientist with an abductive talent through which the range of conjecture is so positioned that retroductive reduction can do efficient work. In operation, this tendency is akin to an animal instinct—but a cognitive instinct that equips the scientist with an at least grossly trustworthy means for discriminating between live options and uselessly wild hypotheses. In one of his Lowell Lectures, Peirce put it this way:

> How was it that man was ever led to entertain . . . [a] true theory? You cannot say that it happened by chance, because the possible theories, if not strictly innumerable, at any rate exceed a trillion—or the third power of a million; and therefore the chances are too overwhelmingly against the single true theory in the twenty or thirty thousand years during which man has been a thinking animal, ever having come into any man's head. Besides, you cannot seriously think that every little chicken, that is hatched, has to rummage through all possible theories until it lights upon the good idea of picking up something and eating it. On the contrary, you think the chicken has an innate idea of doing this; that is to say, that it can think of this, but has no faculty of thinking anything else. The chicken you say pecks by instinct. But if you are going to think every poor chicken endowed with an innate tendency toward a positive truth, why should you think that to man alone this gift is denied? (CP, 5.591 [1903]; compare 2.177 and 8.223)

Peirce has no patience with people who hesitate to accept the existence of such a talent, viewing them as incapable of maintaining the line between sense and foolishness:

> There are minds to whom every predjudice, every presump-

> tion, seems unfair. It is easy to say what minds these are. They are those who never have known what it is to draw a well-grounded induction, and who imagine that other people's knowledge is as nebulous as their own. That all science rolls upon presumption (not of a formal but of a real kind) is no argument with them, because they cannot imagine that there is anything solid in human knowledge. These are the people who waste their time and money upon perpetual motions and other such rubbish. (CP, 6.424 [1878])

To be sure, the accuracy of this abductive sense of the plausible is very limited: "it is an act of *insight*, although of extremely fallible insight."[79] But this fact does not in Peirce's view derogate from its capacity to serve effectively in a crucially important role; indeed, he insists, "[T]he existence of a natural instinct for truth is, after all, the sheet-anchor of science."[80]

3. What Plausibility Does for Us

With Peirce, plausibility is (1) a tool of the "economy of research" that provides needed guidance for the efficient conduct of inquiry, and (2) a tool of historical explanation that accounts for the (quasi-Darwinian) process of the successful evolution of science and for the actual historical course of scientific progress. The first issue will be the theme of the next chapter. Here we shall focus primarily upon the second.

The number of conceivable hypotheses that can be invoked on any question of scientific explanation is literally infinite. Man, a limited creature of finite capabilities, can only begin to scratch the surface. Given that we can draw only a vanishingly small sample from a massive population, what assurance do we have in thinking that any of the whole gamut of the hypotheses we conjecture is worthy of attention?

The facts suggest that man has an abductive talent for doing well at projecting explanatory hypotheses, in whose absence the whole project of scientific theorizing would soon collapse—and, more importantly, would never have gotten off the ground. But how does this abductive talent for formulating and selecting explanatory hypotheses actually work? Whence do we obtain the abductively requisite guidance?

Peirce takes the stance that it is the recourse to plausibility and plausibility alone that justifies the fiat by which we separate the sheep we deem worthy of consideration from the goats we unceremoniously put aside. Our instinctive insight that some theses are more plausible than others is crucial to the very possibility of scientific progress:

> We shall do better to abandon the whole attempt to learn the truth . . . unless we can trust to the human mind's having such a power of guessing right that before very many hypotheses shall have been tried intelligent guessing may be expected to lead us to one which will support all tests, leaving the vast majority of possible hypotheses unexamined. (CP, 6.530 [c. 1901])

Abductive talent, the scientist's appropriate sense of the plausible, is indeed crucial to Peirce's theory that the truth is what lies at the ultimate *focus imaginarius* of inquiry. Abduction tells us where to shine the beam of inquiry's lamp. There is no point in researching, however carefully, in the wrong spot. Did our abductive sense not reflect a tropism for the truth, the prospect of attaining it would be, in Peirce's view, effectively nil.

4. The Rationale of Plausibility: Evolved Instinct as a Basis

What justifies this resort to plausibility? What is the rational warrant for our abductive "talent," our sense of the plausible? Peirce's answer can be conveyed in a single word: *evolution*.

> If you carefully consider with an unbiassed mind all the circumstances of the early history of science and all the other

> facts bearing on the question . . . I am quite sure that you must be brought to acknowledge that man's mind has a natural adaptation to imagining correct theories of some kinds, and in particular to [sic] correct theories about forces, without some glimmer of which he could not form social ties and consequently could not reproduce his kind. In short, the instincts conducive to assimilation of food, and the instincts conducive to reproduction, must have involved from the beginning certain tendencies to think truly about physics, on the one hand, and about psychics, on the other. It is somehow more than a mere figure of speech to say that nature fecundates the mind of man with ideas which, when those ideas grow up, will resemble their father, Nature. (CP, 5.591 [1903])

The resort to plausibility is no mere appeal to animal propensities. Man's evolutionary adaptation endows the human mind with a kind of functional sympathy for the processes of nature:

> [I]f man's mind has been developed under the influence of those [natural] laws, it is to be expected that he should have a *natural light*, or *light of nature*, or *instinctive insight*, or genius, tending to make him guess those laws aright, or nearly aright. This conclusion is confirmed when we find that every species of animal is endowed with a similar genius. . . . It would be too contrary to analogy to suppose that similar gifts were wanting to man. (CP, 5.604 [1903]; compare 6.476)

The key here is that the scientist as a member of *homo sapiens* is the product of an evolutionary development proceeding within nature:

> We now seem launched upon a boundless ocean of possibilities. We have speculations put forth by the greatest masters of physical theorizing of which we can only say that the mere testing of any one of them would occupy a large company of able mathematicians for their whole lives; and that no one such theory seems to have an antecedent probability of being true that exceeds say one chance in a million. When we theorized about molar dynamics we were guided by our in-

> stincts. *Those instincts had some tendency to be true; because they had been formed under the influence of the very laws that we were investigating*. (CP, 7.508 [c. 1898], italics added)[81]

For Peirce the validation of man's abductive instinct lies in evolution. Under the pressure of evolutionary forces, mind of man has come to be "co-natured" with physical reality:

> It is certain that the only hope of retroductive reasoning ever reaching the truth is that there may be some natural tendency toward an agreement between the ideas which suggest themselves to the human mind and those which are concerned in the laws of nature. (CP, 1.81 [c. 1896]; compare 7.220)

The rationale for the reliability of our abductive instincts lies in certain metaphysical convictions whose warrant derives from the cumulative experience of the species:

> [I]nquiry must proceed upon the virtual assumption of sundry logical and metaphysical beliefs; and it is rational to settle the validity of those before undertaking an operation that supposes their truth. Now whether the truth of them be explicitly laid down on critical [i.e., Kantian] grounds, or the doctrine of Common-Sense prevent our pretending to doubt it [is immaterial] . . . . [These] beliefs that appear to be indubitable have the same sort of basis as scientific results have. That is to say, they rest on experience—on the total everyday experience of many generations of multitudinous populations. Such experience is worthless for distinctively scientific purposes, because it does not make the minute distinctions with which science is chiefly concerned . . . although all science, without being aware of it, virtually supposes the truth of the vague results of uncontrolled thought upon such experiences, cannot help doing so, and would have to shut up shop if she should manage to escape accepting them. . . . [T]he instinctive result of human experience ought to have so vastly more weight than any scientific result, that to make laboratory experiments to ascertain, for example, whether there be any uniformity in nature or no,

> would vie with adding a teaspoonful of saccharine to the ocean in order to sweeten it. (CP, 5.521–522 [c. 1905])

Peirce thus treats the "sundry logical and metaphysical beliefs" on which the procedures of science are ultimately founded as being based upon "the instinctive result of human experience."[82]

There is no purely theoretical reason why a hypothesis generated by the abductive procedure should be more promising than purely random guesswork: their standing rests on an appeal to experience. Somewhere along the way, the abductive process will (not "must") yield the correct hypothesis. The very existence of undisputed scientific knowledge was seen by Peirce as a testimony to the truth of this claim, thus making it a scientific, empirical claim about the scientific method itself.[83]

This inductively grounded claim about induction rests on the most fundamental abduction of all: the hypothesis that we can in fact succeed in our attempts at explaining the phenomena of nature:

> Underlying all such principles there is a fundamental and primary abduction, a hypothesis which we must embrace at the outset, however destitute of evidentiary support it may be. That hypothesis is that the facts in hand admit of rationalization, and of rationalization by us. That we must hope they do, for the same reason that a general who has to capture a position or see his country ruined, must go on the hypothesis that there is some way in which he can and shall capture it. We must be animated by that hope concerning the problem we have in hand, whether we extend it to a general postulate covering all facts, or not. Now, that the matter of no new truth can come from induction or from deduction, we have seen. It can only come from abduction; and abduction is, after all, nothing but guessing. We are therefore bound to hope that, although the possible explanations of our facts may be strictly innumerable, yet our mind will be able, in some finite number of guesses, to guess the sole true explanation of them. *That* we are bound to assume, independently of any evidence that it is true. (CP, 7.219 [c. 1901])

Our initial reliance on abduction—prior to its acquisition of experiential credentials—thus rests on a this-or-nothing argument:

> [Retroduction] sets out with a theory and it measures the degree of concordance of that theory with fact. It never can originate any idea whatever. No more can deduction. All the ideas of science come to it by the way of Abduction. Abduction consists in studying facts and devising a theory to explain them. Its only justification is that if we are ever to understand things at all, it must be in that way. (CP, 5.145 [1903])[84]

The idea of a co-naturing of man's mind with physical reality serves a dual function. On the one hand it explains the historically established success of the inductive methods of science. On the other hand it justifies the scientists' reliance on judgments of plausibility. What is at issue is indeed a circular argument "arguing for a scientific hypothesis to support the validity of scientific reasoning" as one critic has put it.[85] But the circle is not vicious. If the efficacy of scientific reasoning is indeed to count as an established fact, we should certainly expect to have a scientific account of it. Science must surely be self-substantiating: our science-provided picture of the world must surely be such that our science-generating inquiry methods turn out to be effective on its telling.[86] (To be sure, this sort of self-substantiation of scientific method should not prove the only substantiation it admits. But Peirce's resort to self-correctiveness as the crucial justificatory consideration puts him on perfectly safe ground here.)

PART TWO: PEIRCE, POPPER, AND THE METHODOLOGICAL TURN

5. Popper's Evolutionary Epistemology

One may well ask, "Just why should the long-term acceptance of a thesis, its mere ultimate survival in the community of scientific inquirers, betoken the *truth* of a thesis?'' The only convincing line of reply takes ''survival'' to mean survival of tests; the thesis has successfully frustrated all experiments or observations designed to prove it false. Peirce's view of survival as probatively truth-indicative results from a proto-Popperian conception that science is committed to efforts to falsify accepted theses, so that the survival of a factual thesis over time becomes an indicator of its acceptability.[87] The difficulty of this approach is that the range of mutually incompatible possibilities that remain unfalsified is always too large for survival-to-date to be unproblematically truth-indicative. When theses or theories can be produced faster than they can be tested and refuted, the fact that a thesis stands as so-far-unrefuted does precious little toward establishing its claim to truth. The key role of abductive talent in Peirce's theory is to aid in resolving difficulties of just this sort.

It will help greatly to clarify Peirce's thinking on these issues of evolutionary epistemology to compare them in more detail with those of Popper.

K. R. Popper's *Objective Knowledge* (Oxford, 1972) presents one of the most fully developed and influential modern versions of epistemological Darwinism in the evolutionary model of scientific progress. Popper's cognitive Darwinism addresses itself specifically to *theories and hypotheses*, approached from the standpoint of their ''fitness to survive by standing up to tests'' (p. 19). Hypotheses arise as variant answers in the context posed by problem-questions. The testing of these hypotheses with a view to their falsification provides a process for ''selection'' among them. The basic idea of Popperian hypothesis-evolution calls for such a mechanism of cognitive variation and selection by ''the method of trial and the elimination of errors'' (p. 70).[88]

The dynamic of this proposed evolutionary process is a cyclic pattern of movement: initial problem to tentative theory to error-elimination to refined problem to refined tentative theory, and so on. ''The neo-Darwinist theory of evolution is assumed; but it is restated by pointing out that its 'mutations' may be interpreted as more or less accidental trial-and-error gambits, and 'natural selection' as one way of controlling them by error-elimination'' (p. 242). The trial and error search procedure at issue here is blind and virtually random. According to Popper, the difference between Einstein and an amoeba is, from the epistemological standpoint, a matter of degree rather than kind, since ''their methods of almost random or cloud-like trial and error movements are fundamentally not very different'' (p. 247). The crucial difference between them lies in the sphere of reactions to solutions: unlike the amoeba, Einstein ''approached his own solutions *critically*'' (p. 247) and subjected them to deliberate falsifying tests. As a result of this eliminative selection of hypotheses, ''our knowledge consists, at every moment, of those hypotheses which have shown their (comparative) fitness by surviving so far on their struggle for existence; a competitive struggle which eliminates those hypotheses which are unfit'' (p. 261).[89]

The model of scientific inquiry presented by Popper rests on a combination of its three basic commitments:

1. With respect to any given scientific issue the number of alternative hypotheses is always in principle infinite.
2. Science proceeds by the tiral and error elimination of hypotheses.
3. This elimination process is inductively blind: Man has no inductive capability for discriminating good from bad hypotheses—of separating the promising from the unpromising, the inherently more plausible from the inherently less plausible—and there is never any reason to think that those hypotheses that have been proposed or considered are somehow more advantageous than those that have not. At every stage, our search among the alternatives must be a matter of blind, random groping.

But now unfortunate consequences loom. The moment we put these premisses together, we destroy any prospect of understanding the success of man's cognitive efforts. The whole achievement of science, its historically demonstrated ability to do its work well and to produce results that if not true are in some way reasonably close to the truth, becomes altogether unintelligible. Indeed, science becomes an accident of virtually miraculous proportions, every bit as fortuitous as someone's correctly guessing at random the telephone numbers of someone else's friends.

Popper is deeply (and apparently proudly) committed to this consequence of his theory. For him the success of science is something fortuitous, accidental, literally miraculous (p. 204) and totally unintelligible:

> However, even on the assumption (which I share) that our quest for knowledge has been very successful so far, and that we now know something of our universe, this success becomes miraculously improbable, and therefore inexplicable; for an appeal to an endless series of improbable accidents is not an explanation. (The best we can do, I suppose, is to investigate the almost incredible evolutionary history of these accidents, from the making of the elements to the making of the organisms.)[90]

On the premisses of Popper's theory of science, the question of why science is as successful as it is—the inquiry after an account of the nature of the world and the nature of man's cognitive technology that can *explain* why our endeavors to acquire knowledge are so effective—must be met with blank *ignorabimus* of intrinsic mystery.

But, as Popper himself stresses (pp. 29–30), discovery of *truth* is the regulative ideal of the process of inquiry. How, then, can the process of Popperian error-elimination ever provide any sort of warrant for our conviction that the actual course of our efforts at inquiry involves a movement, however slow or hesitating, toward this ideal of truth? "We test for truth," Popper maintains, "by eliminating falsehood." But clearly this would work only in the context of a theory of limited possibilities. (We can eliminate endless possibilities as solutions to a problem—say all the odd integers as answers to a Diophantine problem whose real answer is eight—without thereby moving significantly closer to the truth.) Once we grant (as Popper time and again insists) that any hypotheses we may actually entertain are but a few fish drawn from an infinite ocean—are only isolated instances of those infinitely many available hypotheses we have not even entertained, none of which are *prima facie* less meritorious than those we have[91]—then the whole idea of seeking truth by elimination of error becomes pointless. If infinitely many distinct roads issue from the present spot, there is no reason to think that, by elimination one or two (or *n*) of these, we come one jot closer to finding the one that leads to the desired destination.

All this, of course, holds only on the quixotically democratic view that all the possible hypotheses stand on an equal footing, that our process of selection is in no way shrewd but rather is virtually random, that we are not to take the stance that those hypotheses we propose to treat seriously can reasonably be supposed to be more promising candidates than the rest. In short, we are to insist on refusing to credit the human intellect with any inductive skill, any capacity to single out those alternative hypotheses that are (likely to prove) more promising candidates than the rest.

If we are not entitled to regard hypothesis-elimination as narrowing the field of the *real* possibilities, however, this entire eliminative process becomes probatively pointless. The technique of error-elimination is capable of serving the Popperian desideratum of leading closer to the truth only if we are willing to take the un-Popperian step of crediting man's intelligence with Peirce's ''abductive'' capacity for doing reasonably well in selecting hypotheses for testing; if we see inquiry as not constrained to operate by blind trial and error.

Popper explicitly and emphatically insists that ''no theory of knowledge should attempt to explain why we are successful in our attempt to explain things'' (p. 23). And yet this self-denying ordinance is nowhere defended as being inexorably inevitable in the nature of things—the only sort of defense, surely, that could force us to accept so unpalatable and counterintuitive a doctrine. It is difficult to exaggerate the unsatisfactory nature of this position. It fails the basic and rudimentary task of any adequate explanatory theory, that of ''saving the phenomena'' by providing some promising account for them.

An adequate philosophical theory of the rationale of our scientific knowledge of the world must combine a theory of nature and a theory of inquiry in such a way that an account of the success of science is a straightforward and natural result once they are conjoined. But the very best Popper can offer is the thought that our efforts to acquire information about the world by our investigative processes may possibly succeed: "That we cannot give a justification . . . for our guesses [i.e., scientific hypotheses and theories] does not mean we may not have guessed the truth, some of our hypotheses may well be true" (p. 30). Regarding the capacity of scientific inquiry to afford a true picture of reality: "it is not irrational to hope as long as we live—and actions and decisions are constantly forced on us" (p. 101).[92] In the search for truth, finding something that "may well be true" deserves no celebration; it smacks of failure rather than success. Clearly "a not irrational hope" in the adequacy of science is not good enough: what is wanted is a rationally based expectation. If we do not demand an airtight guarantee, we want at least a reasonable assurance that in taking the scientific route to the solution of cognitive problems in the factual area we are doing as well as is possible.

The model of the growth of scientific knowledge along Popperian lines—through the falsification of hypotheses arrived at by blind trial and error—is thus crucially deficient; it is admittedly unable to account for the *reality*, let alone the *rate* of scientific progress. Yet this very issue of the rate and structure of scientific progress is certainly among the basic phenomena that any adequate account of scientific knowledge must be able to explain. Any theory that insists that it is necessary to avoid this issue blazons forth its own inadequacy.

Consider a comparable situation. The vitalistic opponents of strict Darwinism have traditionally objected that evolution has proceeded too quickly and unerringly in devising such highly survival-efficient instruments as, for example, the human eye. They deny that the developmental process could have been wholly the result of natural selection working on random variation. Accordingly, beginning with the "creative evolutionism" of Bergson, vitalists have always objected that the random-variation-cum-natural selection model of the evolutionary process does not provide an adequate account for the rapidity of evolution, and they maintained that the operation of some sort of vital principle is needed to pull the evolutionary process in the right direction and at the right speed. In the case of biological evolution, this objection is doubtless untenable. All the evidence indicates that the available timespan is large enough for the neo-Darwinian mechanisms of mutation and generic selection to do their work. But the case of theory evolution in science is different. There are just too many imaginable hypotheses to be gone through in an entirely inductive blind trial and error search. If our only investigative resource were of this character, then it would indeed have required something verging on a preestablished harmony between scientific guesswork and the ways of nature for us to have come as far as we have over so short a course of human history.[93]

Popper thus faces a vitiating dilemma: he must choose between having the Darwinian selection process operate between all conceivable (*theoretically* available) theories or between all proposed (*actually* espoused) theories. If he opts for the second course, taking the (intrinsically surely attractive and plausible) line that Darwinian selection operates with respect to the actually proposed and genuinely espoused alternatives, then the difficulty of accounting for substantial progress within a limited timespan can be averted only if a capacity for efficiency in hypothesis-conjecture is granted, only if man is conceded a kind of inductive skill, so that those hypotheses actually conjectured are in fact likely to prove among the intrinsically superior alternatives. (As the example of cryptanalysis shows, a shrewd insight into principles of regulative regularity can cut down to reasonable proportions search times that would require astronomical periods on a random trial-and-error basis.) Popper is, of course, emphatically unwilling to concede any such inductive talent for superior hypothesis-conjecture, given his well-known antipathy to anything of inductive-confirmationist tendencies. Consequently, he is driven onto the other horn of the dilemma: his trial-and-error mechanism is saddled with having to grapple with the whole gamut of conceivable alternatives, becoming trapped in the problem of time-availability and unrationalizable rates of progress.

6. The Role of Trial and Error

This critique of Popperian evolutionism by a blindly groping trial and error development of theories was anticipated by Peirce not only in its general tendency but in its details. Peirce insists that trial and error cannot adequately account for the existing facts and that man's intellect must be credited with a truth-tropism.

> [T]ruths, on the average, have a greater tendency to get believed than falsities have. Were it otherwise, considering that there are myriads of false hypotheses to account for any given phenomenon, against one sole true one (or if you will have it so, against every true one), the first step towards genuine knowledge must have been next door to a miracle. (CP, 5.431 [1905])

Peirce's detailed arguments in this trend merit quotation at considerable length:

> But how is it that all this truth has ever been lit up by a process in which there is no compulsiveness nor tendency toward compulsiveness? Is it by chance? Consider the multitude of theories that might have been suggested. A physicist comes across some new phenomenon in his laboratory. How does he know but the conjunctions of the planets have something to do with it or that it is not perhaps because the dowager empress of China has at that same time a year ago chanced to pronounce some word of mystical power or some invisible jinnee may be present. Think of what trillions of trillions of hypotheses might be made of which one only is true; and yet after two or three or at the very most a dozen guesses, the physicist hits pretty nearly on the correct hypothesis. By chance he would not have been likely to do so in the whole time that has elapsed since the earth was solidified. You may tell me that astrological and magical hypotheses were resorted to at first and that it is only by degrees that we have learned certain general laws of nature in consequence of which the physicist seeks for the explanation of

his phenomenon within the four walls of his laboratory. But when you look at the matter more narrowly, the matter is not to be accounted for in any considerable measure in that way. Take a broad view of the matter. Man has not been engaged upon scientific problems for over twenty thousand years or so. But put it at ten times that if you like. But that is not a hundred thousandth part of the time that he might have been expected to have been searching for his first scientific theory.

You may produce this or that excellent psychological account of the matter. But let me tell you that all the psychology in the world will leave the logical problem just where it was. I might occupy hours in developing that point. I must pass it by.

You may say that evolution accounts for one thing. I don't doubt it is evolution. But as for explaining evolution by chance, there has not been time enough.

However man may have acquired his faculty of divining the ways of nature, it has certainly not been by a self-controlled and critical logic. Even now he cannot give any exact reason for his best guesses. It appears to me that the clearest statement we can make of the logical situation—the freest from all questionable admixture—is to say that man has a certain insight, not strong enough to be oftener right than wrong, but strong enough not to be overwhelmingly more often wrong than right, into . . . nature. An insight, I call it, because it is to be referred to the same general class of operations to which perceptive judgments belong. This faculty is at the same time of the general nature of instinct, resembling the instincts of the animals in its so far surpassing the general powers of our reason and for its directing us as if we were in possession of facts that are entirely beyond the reach of our senses. It resembles instinct too in its small liability to error; for though it goes wrong oftener than right, yet the relative frequency with which it is right is on the whole the most wonderful thing in our constitution. (CP, 5.172–173 [1903])[94]

In another passage, Peirce develops this train of ideas further:

[Think of] those considerations which tend toward an expectation that a given hypothesis may be true. These are of two kinds, the purely instinctive and the reasoned. In regard to instinctive considerations, I have already pointed out that it is a primary hypothesis underlying all abduction that the human mind is akin to the truth in the sense that in a finite number of guesses it will light upon the correct hypothesis. . . . From the instinctive, we pass to reasoned, marks of truth in the hypothesis. . . . The game of twenty questions is instructive. In this game, one party thinks of some individual object, real or fictitious, which is well-known to all educated people. The other party is entitled to answers to any twenty interrogatories they propound which can be answered by *Yes* or *No,* and are then to guess what was thought of, if they can. If the questioning is skillful, the object will invariably be guessed; but if the questioners allow themselves to be led astray by the will-o-the-wisp of any prepossession, they will almost as infallibly come to grief. The uniform success of good questioners is based upon the circumstance that the entire collection of individual objects well-known to all the world does not amount to a million. If, therefore, each question could exactly bisect the possibilities, so that *yes* and *no* were equally probable, the right object would be identified among a collection numbering 2^{20}. Now the logarithm of 2 being 0.30103, that of its twentieth power is 6.0206, which is the logarithm of about 1,000,000 $(1 + .02 \times 2.3)(1 + .0006 \times 2.3)$ or over one million and forty-seven thousand, or more than the entire number of objects from which the selection has been made. Thus, twenty skillful hypotheses will ascertain what two hundred thousand stupid ones might fail to do. The secret of the business lies in the caution which breaks a hypothesis up into its smallest logical components, and only risks one of them at a time. What a world of futile controversy and of confused experimentation might have been saved if this principle had guided investigations . . . (CP, 7.220 [c. 1901])

With magisterial shrewdness Peirce puts his finger upon exactly the right point: an evolutionary model of random trial and error with respect to possible hypotheses just cannot operate adequately within the actual (or perhaps even any possible) timespan.[95] His diagnosis of the difficulty is excellent. But what of his proposed cure?

As we have seen, his own response is a theory of an abductive talent based upon a cognitive instinct evolved over countless millennia of interaction between mind and the wider domain of nature. Such a resort to instinct is clearly a philosophically problematic expedient. Is there a more palatable alternative?

7. Instinctival versus Methodological Darwinism

The one and only point on which it seems advisable to part company with Peirce in this matter is to explicitly and deliberately substitute the *methodology* of inquiry and substantiation for his somewhat mysterious capacity of *insight* or *instinct*. On this revised approach, the venture of hypothesis formulation is guided by heuristic principles of method, involving the use of proven methods which themselves have eventually emerged from a process of trial-and-error in inquiry. It seems reasonable to take the stance that scientific discovery (or at any rate scientific conjecture) is subject to the guidance of rational principles of search, principles whose basis is a matter of method and not instinct. We thus retain the Peircean recourse to plausibility, but alter drastically the basis on which the plausibility rests.

Hypotheses are not created *ex nihilo* by random groping; they are constructed on a suitable methodological foundation. They emerge not from random combinations but from the detection of patterns in the empirical data. Without such methodological guidance, we are driven to a "method" that is in effect the absence of method, the "method of last resort" as it were, a merely random search through the possibilities. It is just at this point that methodological, regulative, and procedural considerations can come into effective operation.[96] We cannot at each stage of inquiry place the whole spectrum of logically possible alternatives on equal footing. Of course, we cannot eliminate the "implausible" candidates on the basis of *certain* knowledge, since the operative principles of analogy and coherence are only *presumptive* in their force. Such a cognitively constitutive stance is not appropriate. But we can take the cognitively regulative approach that certain sorts of alternatives ("plausible" on the basis of preserving analogies) can be taken as more worthy of serious considerations.

In effect, the shift from instincts to methods (that is, methods for thesis-substantiation) enables us to have it both ways. We avoid occultism by relying at the methodological level on a strictly trial and error mechanism of learning; and we avoid the rational impotence of not being able to account for the actual course of scientific progress. The combination of a model of method-learning, based on the blind groping of trial and error, and of thesis-learning, based on the use of methods, makes it possible to have the best of both worlds.[97] We can accept a model of cognitive progress based on the mechanism of pure trial and error, but we reorient its applicability from theses (theories) towards methods for thesis-selection. We also are able to account for the rapidity of scientific progress in a straightforward methodological way. Our approach thus proceeds by inserting the matter of methodology as a mediating link emplaced between the operation of a trial and error mechanism and the espousal of factual theses.

A possible objection looms. The process of rational selection implements an evolutionary model in which substantive adequacy is correlative with historical survival. But, of course, people are not as rational as that; they have their moments of aberration and folly. Might not such tendencies selectively favor the survival of the fallacious rather than the true? Might not the process of cognitive evolution be shunted away from truth and our sense of the plausible distorted? Peirce certainly recognized this prospect:

> Logicality in regard to practical matters . . . is the most useful quality an animal can possess, and might, therefore, result from the action of natural selection; but outside of these it is

> probably of more advantage to the animal to have his mind filled with pleasing and encouraging visions, independently of their truth; and thus, upon unpractical subjects, natural selection might occasion a fallacious tendency of thought. (CP, 5.366 [c. 1877])[98]

However, the methodological and regulative orientation of our present approach to plausibility again safeguards against such a fallacious tendency. At the level of individual beliefs, ''pleasing and encouraging visions'' might indeed receive a survival-favoring impetus; but this unpleasant prospect is effectively removed where a systematic method of inquiry is concerned. This method must by its very synoptic nature penetrate deeply throughout the sphere of the pragmatically effective.

In sum, these considerations suggest a possible and promising alternative to Peirce's handling of the key issue of abductive talent. For this can now be construed not as a matter of a historically evolved instinct for creating truth-approximating hypotheses, but rather as a historically developed methodology for guiding the search for efficiently data-accommodating hypotheses. Such an approach envisages a transformation from an instinct for hypothesis selection into a ''logic'' or methodological organon for hypothesis-construction.[99] This shift from an instinctive and biological to a methodological and cultural hypothesis-selection process enables us to preserve all the advantages of Peirce's approach, while avoiding its problematic reliance on a somewhat mysterious instinct.[100]

4 Peirce and the Economy of Research

1. Peirce's Project

In his analysis of scientific method, Peirce gave the place of pride to a theory—indeed a discipline—of his own devising. He called it the economy of research.[101] This Peircean project, unfortunately, has been grossly neglected by his commentators and, even more regrettably, by the subsequent course of development in the philosophy of science.

As we have already seen, Peirce divided the labor of hypothetico-inductive inquiry in science between two distinct processes or procedures. The first, *abduction*, has to do with the elaboration of possible (but plausible) hypotheses and the provision of possible (but promising) explanations for the solution of scientific problems. The second, *retroduction*, is concerned with narrowing this range of alternative possibilities to the one that is in fact correct, or at any rate, appears the optimal candidate for correctness under the epistemic circumstances.[102] Peirce was acutely alive to what has subsequently become a commonplace in the philosophy of science, the idea that data underdetermine theory because an infinitude of alternative hypotheses will always conform equally well to any finite body of empirical data. If science is to proceed in its quest for adequate theories by testing hypotheses against the naturally available or experimentally contrived empirical facts, it must seek preexperiential guidance to determine priorities as it works through the perplexing wealth of alternative possibilities. The crucial first step of the inductive process of hypothesis testing will thus be to decide which of the hypotheses that are in principle available merit immediate checking, which can be put off until tomorrow, and which can wait until Niagara runs dry.

Peirce holds that it is to the economy of research that we are to look for guidance here.[103] For whatever intrinsic appeal may be possessed by various individual items within the welter of abductively eligible hypotheses, we must face up to the harsh time and material limitations imposed by the cost of experimental research:

> [I]f two hypotheses present themselves, one of which can be satisfactorily tested in two or three days, while the testing of the other might occupy a month, the former should be tried first, even if its apparent likelihood is a good deal less. . . . In an extreme case, where the likelihood is of an unmistakably objective character, and is strongly supported by good inductions, I would allow it to cause the [indefinite] postponement of the testing of a hypothesis. For example, if a man came to me and pretended to be able to turn lead into gold, I should say to him, "My dear sir, I haven't time to make gold." But even then the likelihood would not weigh with me directly, as such, but because it would become a factor in what really is in all cases the leading consideration in Abduction, which is the question of Economy—Economy of money, time, thought, and energy. (CP, 5.598–600 [1903]: compare 6.528ff.)

Thus the methodology of inductive practice is, in Peirce's view, pivotally dependent on the intelligent deployment of economic considerations from the very outset. For when it comes to the crucial process of checking and testing, we can deal with only a small number of possibilities, given limited time and limited resources. Owing to the economic exigencies of our situation, most of our candidate hypotheses must simply be put aside untested and even generally unconsidered. The selection of hypotheses for testing must be determined on a strictly economic basis:

> Proposals for hypotheses inundate us in an overwhelming

> flood, while the process of verification to which each one must be subjected before it can count as at all an item, even of likely knowledge, is so very costly in time, energy, and money—and consequently in ideas which might have been had for that time, energy, and money, that Economy would override every other consideration even if there were any other serious considerations. In fact there are no others. (CP, 5.602 [1903])[104]

Once abductive conjectures lie before us, considerations of economy become the ruling factor in inductive inquiry:

> Let us suppose that there are thirty-two different possible ways of explaining a set of phenomena. Then, thirty-one hypotheses must be rejected. . . . Now the testing of a hypothesis is usually more or less costly. Not infrequently the whole life's labor of a number of able men is required to disprove a single hypothesis and get rid of it. Meantime the number of possible hypotheses concerning the truth or falsity of which we really know nothing, or next to nothing, may be very great. In questions of physics there is sometimes an infinite multitude of such possible hypotheses. The question of economy is clearly a very grave one. (CP, 6.529–530 [c. 1901]; compare 2.759)

Peirce thus maintained that the economics of research is a pivotally determinative consideration when selecting hypotheses for scientific testing. He set out the spectrum of relevant considerations in the following terms:

> [I]n view of the fact that the true hypothesis is only one out of innumerable possible false ones, in view, too, of the enormous expensiveness of experimentation in money, time, energy, and thought . . . the consideration of economy [is critical]. Now economy . . . depends upon three kinds of factors: cost; the value of the thing proposed, in itself; and its effect upon other projects. Under the head of cost, if a hypothesis can be put to the test of experiment with very little expense of any kind, that should be regarded as a recommendation for giving it precedence in the inductive procedure. For even if it be barely ad-

> missible for other reasons, still it may clear the ground to have disposed of it. In the beginning of the wonderful reasonings by which the cuneiform inscriptions were made legible, one or two hypotheses which were never considered likely were taken up and soon refuted with great advantage. (CP, 7.220 [c. 1901])

Peirce went on to elaborate various other aspects of abductive procedure:

> In those subjects, we may, with great confidence, follow the rule that that one of all admissible hypotheses which seems the simplest to the human mind ought to be taken up for examination first. Perhaps we cannot do better than to extend this rule to all subjects where a very simple hypothesis is at all admissible.
>
> This rule has another advantage, which is that the simplest hypotheses are those of which the consequences are most readily deduced and compared with observation; so that, if they are wrong, they can be eliminated at less expense than any others.
>
> This remark at once suggests another rule, namely, that if there be any hypothesis which we happen to be well provided with means for testing, or which, for any reason, promises not to detain us long, unless it be true, that hypothesis ought to be taken up early for examination. Sometimes, the very fact that a hypothesis is improbable recommends it for provisional acceptance on probation.
>
> On the other hand, if one of the admissible hypotheses presents a marked probability of the nature of an objective fact, it may in the long run promote economy to give it an early trial. . . . Such a probability must be distinguished from a mere likelihood which is nothing better than the expression of our preconceived ideas. (CP, 6.532–534 [c. 1910])

Considerations of the sheer "tactics of research," in the sense of the efficiency of search procedures, are also allowed to enter in:

> There still remains one more economic consideration in refer-

> ence to a hypothesis; namely, that it may give a good "leave," as the billiard-players say. If it does not suit the facts, still the comparison with the facts may be instructive with reference to the next hypothesis. (CP, 7.221 [c. 1901])

Peirce proposed to construe the economic process at issue as the sort of balance of assets and liabilities that we today would call cost-benefit analysis.[105] On the side of benefits, he was prepared to consider a wide variety of factors: closeness of fit to data, explanatory value, novelty, simplicity, accuracy of detail,[106] precision, parsimony,[107] concordance with other accepted theories, even antecedent likelihood and intuitive appeal.[108] But in the liability column, there sit those hard-faced factors of "the dismal science": time, effort, energy, and last but not least, crass old money.[109]

It is worth stressing that the abductive criteria of hypothesis selection need not correlate with the probative criteria of hypothesis adoption upon completion of successful retroductive tests. To be sure, all of the latter must be included among the former ("closeness of fit to data," intertheoretic concordance, simplicity, etc.), but as Peirce sees it the reverse is not true. There will be some abductively effective parameters (novelty and intuitive appeal, for example) that lack any probative force. Certain criteria that militate on behalf of considering a hypothesis will fail to represent criteria for simply accepting it as presumably true. Moreover, adoption and implementation may well pose economic problems of their own, issues that go beyond the economic aspects of abductive inquiry. The economics of research is a complex matter.

The availability of resources is, for Peirce, one of the pivotal determinants of the course of scientific inquiry:

> Suppose that to avoid wasting a great deal of time upon a hypothesis which the first comparisons with the facts may show to be utterly worthless, an investigator of a certain conjecture draws up and resolves to follow a well-considered initial program for work upon the question, and that this consists mainly in working out and testing as many consequences of the hypothesis as he can work out by a certain mathematical

> method and can ascertain the truth or falsity of at a cost of not more than $100 for each. . . . The predictions must *eventually* be so varied as to test every feature of the hypothesis; yet the interests of science command constant attention to economy, especially in the earlier inductive stages of research. (CP, 2.759 [c. 1905])

Peirce formulated the task of the economy of research in words that can scarcely be improved upon a century later:

> The doctrine of economy, in general, treats of the relations between utility and cost. That branch of it which relates to research considers the relations between the utility and the cost of diminishing [the gaps and] the probable error of our knowledge. Its main problem is, how, with a given expenditure of money, time, and energy, to obtain the most valuable addition to our knowledge. (CP, 7.140 [c. 1879])

His commendation for this enterprise is unequivocal: "that economical science is particularly profitable to science; and that of all branches of economy, the economy of research is perhaps the most profitable . . . costing little beyond the energies of the researcher, and helping the economy of every other science" (CP, 7.161 [1879]).

One of the main advantages of an economic approach in this domain is its potential savings of effort and resources in the face of the operation of a principle of diminishing returns in inquiry, a conception to which Peirce returned again and again.[110] Indeed, he tended to view the uncovering of a law of diminishing returns as his most important discovery in the "economy of research." According to this law, the expected marginal utility of further investigation decreases as experimentation continues. Here Peirce took the sorts of mensurational enterprise that he was familiar with in his work for the Coast and Geodetic Survey as typical of scientific investigations, and based his analysis on his experience of probable errors of estimated quantities as convex functions of sample size.[111] His unfortunate choice of such a paradigm led him to espouse a geographic exploration model of scientific progress along the lines already criticized in chapter 2 above. In this particular connection, his economic approach did not serve him well.

But, more fundamentally, considerations of economy are the pivot point around which the whole operation of rational inquiry will have to turn. As we have seen, Peirce held that the inductive approach must ultimately yield truth even in the hands of the most bumbling of determined inquirers:

> It is true that, however carelessly the abduction is performed, the true hypothesis will get suggested at last. But the aid which a correct logic can afford to science consists in enabling that to be done at small expenditure of every kind which, at any rate, is bound to get done somehow. The whole service of logic to science, whatever the nature of its services to individuals may be, is of the nature of an economy. (CP, 7.220, fn. 18 [c. 1901])

The task of a rational, economically sagacious methodology of inquiry is to expedite this effectively inevitable result:

> Everything thus depends upon rational methods of inquiry. They will make that result as speedy as possible, which otherwise would have kicked its heels in the anteroom of chance. Let us remember, then, that the precise practical service of sound theory of logic is to abbreviate the time of waiting to know the truth, to expedite the predestined result. But I here use the words "abbreviate" and "expedite" in a peculiar sense. Imagine a derelict wreck to be floating about on the ocean; and suppose that it will be driven hither and thither until it chances to be cast upon a shore. Then, a vessel which should go and take that derelict in tow and deliberately strand it upon the nearest shore, would be "abbreviating" or "expediting" the fulfillment of the destiny of that derelict in the same sense in which I hold that logic "abbreviates" inquiry, and "expedites" its result. It changes a fortuitous event which may take weeks or may take many decennia into an operation governed by intelligence, which will be finished within a month. This is

> the sense in which logic "abbreviates" and "expedites" the attainment of truth. (CP, 7.78 [c. 1902])

The realization of efficiency, whose cultivation is after all the very *raison d'être* of the economy of research, is the governing consideration for that "sound theory of logic" whose articulation is the aim of Peirce's theory of scientific method.

The introduction of an economic perspective does not detract from the value of science as an intrinsically interesting venture with a perfectly valid *l'art pour l'art* aspect. As Peirce sees it, it simply recognizes the inevitable economic aspect of any human enterprise, however great may be the purely cognitive, theory-oriented and "transcendental" values at issue:

> The value of knowledge is, for the purposes of science, in one sense absolute. It is not to be measured, it may be said, in money; in one sense that is true. But knowledge that leads to other knowledge is more valuable in proportion to the trouble it saves in the way of expenditure to get that other knowledge. Having a certain fund of energy, time, money, etc., all of which are merchantable articles to spend upon research, the question is how much is to be allowed to each investigation; and *for us* the value of that investigation is the amount of money it will pay us to spend upon it. *Relatively*, therefore, knowledge, even of a purely scientific kind, has a money value. (CP, 1.122 [c. 1896])

2. The Unmerited Neglect of Peirce's Project

To this idea of the economy of research—of cost-benefit analysis in inductive inquiry and reasoning—Peirce gave as central a place in his methodology of science as words can manage to assign.[112] Yet no other part of this great man's philosophizing has fallen on stonier ground. The Peirce bibliographies now run to over eight hundred entries and currently add over forty items annually.[113] Yet in all this flood of writing, extending over the century since Peirce flourished,[114] there is not one single item that devotes substantial attention to the analysis of this aspect of his theory of science.[115] This is a great misfortune, for there is virtually no part of Peirce's philosophy that is currently more relevant and is capable of rendering greater service to the solution of current disputes.[116] Everything considered, it is no exaggeration to say that Peirce's project of the economy of research is an instrument that can cut through many recent disputes about inductive reasoning like a hot knife through butter. To persuade the reader of this large claim, we shall embark on a rapid guided tour over some of the main controversies of the recent theory of inductive reasoning.

3. Carnap's Requirement of Total Evidence

Let us begin by considering one of the major controverted principles of recent inductive logic, Carnap's requirement of total evidence. Carnap states this requirement as follows: "In the application of inductive logic to a given knowledge situation, the total evidence available must be taken as basis for determining the degree of confirmation."[117] Given a body of (relevant and correct) evidence e the conscientious scientist will never settle for an assessment of the degree of confirmation of a hypothesis h as having the value $dc(h/e)$ as long as there is a further available item of (relevant and correct) evidence e', so that $dc(h/e \ \& \ e')$ would represent the appropriate quantity. Thus the confirmationist's work, like that of the proverbial housewife, is virtually never done.

The point at issue is quite clear in case of statistically based probabilities.[118] Suppose we inquire regarding the chances of X's having cardiac problems of a certain type by a particular age. How far are we to carry the issue of extracting the evidence that is "available" from a suitable reference class? Is this to be Americans in general, American males, male Pennsylvanians, members of X's profession, members of X's family, members of X's family with X's reading habits, etc. Where is the introduction of fuller and more complex information ever to stop? We are thus brought to the standard objection that total evidence is an inherently impracticable demand. There is no theoretical limit to the relevant information that is (in principle) actually "available." Evidential totalitarianism leads ultimately *ad absurdum*, since it carries us to the statistically vitiating conclusion of narrowing our focus of consideration to the single case itself.

The Carnapian total evidence requirement is in fact the signpost of a crucial tension. On the one hand, it would be rationally indefensible to neglect some crucial available datum—*X*'s employment as an asbestos-factory worker, say, or devotion to cigarettes. To be sure of getting enough into the evidence-base, Carnap insists that we put in *everything*. On the other hand, we can hardly be expected to go on an endless goose-chase through the whole gamut of *X*'s characteristic properties, including his penchant for blondes or preference for peaches over pears. It seems somehow neurotic to insist on working literally everything into the account.

The minute we introduce the economic aspect, however, the situation is transformed. On the basis of purely theoretical considerations, we are entrapped in the dilemma that a total evidence requirement is indispensable on the one hand and unworkable on the other. But economic factors will operate to supply the definiteness needed to secure a workable approach to this aspect of the methodology of inductive inquiry. At once it becomes clear that total evidence is not the real issue but rather the maximal volume of duly relevant evidence that can be obtained relative to the available resources (or better, relative to the resources it makes sense to make available considering the intrinsic importance of the problem for which the probability at issue is being determined).[119]

4. Hempel's Paradox of the Ravens

A substantial literature has grown up over a problem of inductive reasoning first posed by C. G. Hempel in 1946, the so-called "Paradox of the Ravens."[120] It is rooted in the observation of deductive logic that "All *X* is *Y*" is equivalent by contraposition to "All $\overline{Y}$ is $\overline{X}$." Accordingly, "All ravens (R) are black (B)" is deductively equivalent to "All non-black-objects ($\bar{B}$) are non-ravens ($\bar{R}$)." Given this equivalence, why should it be that in inductive contexts one is inclined to accept black ravens as confirming instances of the claim but not white tennis shoes?

Regard this situation as depicted in a Venn diagram:

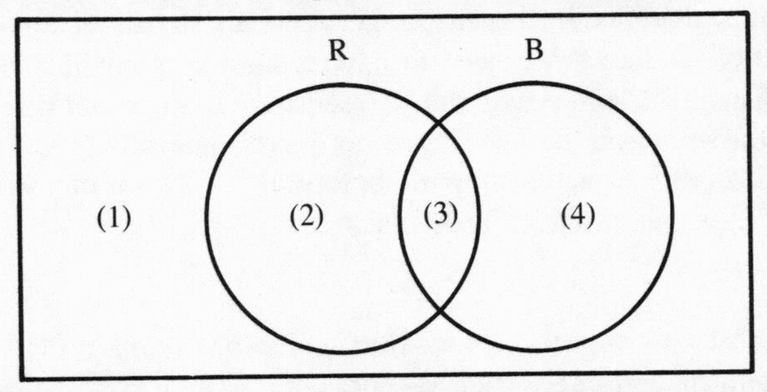

The emptiness of (2) is quite equivalently checked by going to the R's = (2) + (3) and seeing that only (3)'s are encountered as by going to the $\bar{B}$'s = (1) + (2) and seeing that only (1)'s are encountered. Either way, we are simply checking the emptiness of compartment (2). There seems to be no logico-theoretical reason for granting one of these approaches a preferred status over the other.

But there is a crucial economic difference between the two approaches. To check the emptiness of (2) by the "natural" approach (through the R's) means going to the R's = (2) + (3) and checking their color. To establish it by the "unnatural" approach (through the $\bar{B}$) means going to the $\bar{B}$'s = (1) + (2) and checking their type. Now let's make a few assumptions to delineate the structure of the situation:

a. The number of R's is approximately 10^8.
b. The number of $\bar{B}$'s is approximately 10^{40}.
c. The average cost of finding an R is around \$1.
d. The cost of determining the blackness of an R-in-hand is approximately 1¢ = \$.01.
e. The average cost of finding a $\bar{B}$ is around .1¢ = \$.001.
f. The cost of determining the ravenhood of a $\bar{B}$-in-hand is approximately .1¢ = \$.001.

Nothing much hinges on the particular numbers here. Specifically regarding (b), there is little point in contemplating some Eddington-like estimate of the number of molecules in the universe; all that we need for present purposes is that the number of identifiable objects in the world be rather big.

Let it be assumed further that to get adequate statistical control of a population of size X we need to have a sample of roughly the size $\sqrt{X}$. (The result will prove highly insensitive to the exact size of this control assumption; its details are not worth quarreling over.)

Two courses of action now lie before us: approach one is to find $(10^8)^{1/2}$ ravens and check their blackness:

$$\text{Cost: } 10^4 \times \$1.01 = \$10{,}100$$

Contrast now approach two, which calls for finding $(10^{40})^{1/2}$ non-black objects and check their ravenhood:

$$\text{Cost: } 10^{20} \times \$.002 = \$2 \times 10^{17}$$

There is a striking economic difference between these two strategies of verification, and this difference puts black ravens and white tennis shoes on an altogether different plane. If someone hands us a verified black raven, he has contributed one hundredth of one percent to the cost of the whole project of verification. If he hands us the white tennis shoe, his contribution is vanishingly small. The natural course of approach is massively cheaper; with it, expenditure at any given level goes enormously further towards adequacy (that is, buys vastly more by way of confirmation/disconfirmation). Reliance on the "natural" instances wins hands down in cost-effectiveness as an inductive strategy.[121] The Peircean approach to the economic dimension once again renders good service.

5. Goodman's Grue Paradox

Let us now turn to Goodman's "new riddle of induction." In an influential essay of 1953, Nelson Goodman launched a problem for the theory of inductive inference which has occasioned an enormous literature over the ensuing years. Goodman's puzzle is based on a somewhat unorthodox pair of color concepts:

grue = examined before the temporal reference-point t_0 and is green or is not examined before t_0 and is blue. (t_0 is an otherwise arbitrary moment of time that is not in the past.)

bleen = examined before the temporal reference-point t_0 and is blue or not examined before t_0 and is green.

If we do our inductive reasoning on the basis of *this* color taxonomy, we shall seemingly obtain excellent inductive support for the thesis that all emeralds will eventually have the appearance we have standardly indicated by the description "blue" (after t_0)—since all that have been examined to date are grue. On this basis, our "normal" inductive expectations would be totally baffled, and we arrive at a Hume-like result from a totally un-Humean point of departure.[122]

Various considerations mark this issue as an inductive puzzle that cuts deeper than one at first might think.

1. No amount of empirical evidence can help us choose between the "abnormal" grue/bleen color taxonomy and the normal green/blue one. Empirical evidence relates in the nature of things to past-or-present and there is (*ex hypothesi*) no difference here.

2. We cannot object that grue and bleen make explicit reference to time (to t_0), because this only seems so from the parochial standpoint of our familiar color terminology. From the grue/bleen standpoint, the shoe is on the other foot, for, from the perspective of the grue-bleeners, it is *our* color taxonomy that appears to be time-dependent:

green = examined before t_0 and grue or not examined before t_0 and bleen.

blue = examined before t_0 and bleen or not examined before t_0 and grue.

The situation is entirely symmetric from *their* perspective, and what's sauce for the goose is also sauce for the gander as far as any theoretical objections on the basis of general principles go.

The situation thus looks to be one of total parity as between our color talk and that of the grue-bleeners, with no theoretical advantage available to decide the choice between them and us. Goodman himself in effect gives up on finding a preferential rationale of choice on a theoretical basis of general principle and falls back on an appeal to "entrenchment," the fact (in the final analysis) that custom and habit has settled the issue preemptively in favor of our established color-language habits. (Few commentators have found this resolution convincing.)[123]

But the whole issue wears a rather different aspect when approached from the direction of economic considerations. To operate with grue/bleen it would not suffice for the application of one's color taxonomy merely to recall that the thing looked just like *this* or like *that*, if we forgot whether we saw it before or after t_0. Again, it would not serve to have phenomenologically faithful color photographs of the thing if we had no record of when they were taken, and so we could not decide (say) whether one was taken before t_0 and the other after t_0, or whether both were merely different prints of the same picture.

In the orthodox green/blue color taxonomy, all that ever matters—not only with current perception but also with pictorial records, memory, and even precognition—is the strictly phenomenal issue of the perceived appearance of things. In the ordinary case, but not with grue/bleen, it happens that ostension is an adequate apparatus for learning and teaching because phenomenal appearance is the only issue.[124] Thus green/blue can be handled by purely ostensive surrogates (colored just like *this*—pointing to grass—or like *that*—pointing to the sky), while grue/bleen involves the additional issue, the additional complication, of the time of observation relative to the great divide at t_0. The orthodox case involves simply the ostensively manipulable phenomenology of observation; the unorthodox case also calls for the chronometric recourse to temporal data. It is a matter of economy—not of chance!—that the "normal" taxonomy is normal.

These considerations indicate a deep asymmetry between the two cases. The orthodox color taxonomy is in principle cheaper to operate with than the latter as a basis for accommodating our descriptive and, above all, our inductively projecture purposes. In one case color talk can be carried on wholly in an ostensively taught language—where all that matters is the surface appearance of things—in a perfectly uniform way vis-à-vis time. In the other case, we have a two-factor mechanism that adds to this ostensively accessible phenomenology a layer of chronometric issues arising through the areas of learning, teaching, explanation, and application. Thus the orthodox taxonomy is clearly easier and cheaper to operate.

I want to stress that this is not an externalized issue that involves an invidious comparison of one color taxonomy case from the unfairly conceded vantage point of the other. It is a strictly internal issue of what it takes—not just *de facto* but in principle, in "every possible world," so to speak—to operate with the machinery of these rival color taxonomies. We now definitely do not have an invidious comparison of one color taxonomy from the standpoint of a precommitment to the other but simply the question of the resources or mechanisms needed for making the one taxonomy or the other operate effectively on its own, "from within."

We are accordingly enabled to push the issue a step further than Goodman did. We are not left with the brute fact of entrenchment itself—that the orthodox taxonomy is entrenched vis-à-vis its alternatives. Rather, we get some insight into *why* it has become entrenched, namely, that it is in principle less complex and more convenient (economical!) to operate. Entrenchment is thus rendered operative not as a matter of mere historical happenstance but as a factor that obtains a perfectly sound rationale on the basis of economy.[125]

It thus appears that the problem of validating our inductive recourse to the orthodox color taxonomy vis-à-vis its Goodmanian rivals can be resolved on matters of principle rather than "mere facts," such as Hume's "custom" or Goodman's "entrenchment." To be sure, the presently operative principles are practical, that is, economic, rather than theoretical in import. And the considerations needed to cut through the Gordian knot are precisely those involved in the economic factors of efficiency and convenience operative in the Peircean economy of inductive research.

6. Generality-Preference and Falsificationism

Let us now turn to another issue, the significance of relative generality in inductive inquiry. To begin with, it will be clear that generality—simple extent of reach and range—is going to play a central role in any theory of economy of research where cost-effectiveness is a consideration. On the orthodox approach, science is interested in scrutinizing general theses because they are maximally *informative*. On a Popperian approach, science is generality-oriented because general theses are more vulnerable, more falsifiable—optimally *testable*. On a Peircean approach, science is generality-oriented because general theses are the most *cost-effective*.

Consider the sequence:

1. All (10) lions in the zoo have ϕ.
2. All (1,000) lions in the USA have ϕ.
3. All (100,000) living lions have ϕ.
4. All (100,000,000) lions who have ever lived have ϕ.

Let us compare the relative advantages of these generalizations as foci of research. Two considerations will now enter in: First, the (previously supposed) fact that to have reasonably good statistical control over a population of size N we must check a sample of (say) $\sqrt{N}$ of that population (given reasonable randomness of selection). Second, in selecting X individuals of a certain sort, the average cost of the operation decreases with the size of X by a mass-production effect (say that the average unit cost $\propto 1/\log X$, subject to the well-established principle of the economics of production that the cost of the nth unit is proportional to $1/n$). Thus in comparing 1 – 4 above, it becomes clear that we will get a 10,000,000-fold increase in the range of application for a 500-fold increase in expenditure. Generality is obviously very advantageous for cost-effectiveness.

Thus, from the orthodox approach, general hypotheses are the most informative, on a Popperian approach they are the most testable, and on a Peircean approach they are superior because they combine these desiderata: they offer the most content per unit effort to be invested in testing. (The more general the thesis, the larger will be the "cognitive bang for the buck," other things being equal.)

So far so good. But exactly how decisive is generality as a factor in scientific inquiry? In this context let us focus on some of the issues regarding the theories of Karl Popper on the methodology of science—beginning with the Popperian thesis, "It is this interest in the testability of hypotheses which leads . . . to my demand that . . . statements of a high level of universality should be chosen for scrutiny and testing."[126] As Popper sees it, the rational agenda of inquiry calls for science to give priority to the most general (daring and therefore vulnerable) hypotheses in a problem-situation.

The case stands as follows:

Issue: The choice of subset-of-initial-concern among mutually exclusive hypotheses of different levels of generality.

Question: Are we to choose the *less* general (safest and most cautious) or the *more* general (riskiest and most ambitious) has having prime priority-claims to investigation?

Here we confront the question of the defensibility of a generality-preference.

TABLE 1

ASSUMED SITUATION OF TWO GROUPS OF PROBLEM-RESOLVING HYPOTHESES

	Group A	*Group B*
Number	10	100
Probability of each (on the basis of a "best estimate" relative to the evidence-at-hand)	.05	.005
Generality of each*	20	80
Resource cost of testing	10 units	1 unit

*For simplicity we suppose that generality may be ranked on a scale from 0 to 100 (say for the percentage of the theoretically relevant species of individuals to which hypotheses of the range at issue might conceivably apply).

Suppose, for example, that there is a scientific problem-issue for which we have 110 alternative, hypothetically available solutions, falling into two groups as shown in table 1. Suppose further that 120 units of resources are at our disposal. We are to put x hypotheses of group A, and y of group B to the test. How are we to decide on specific values for x and y? If we are true Popperians, we turn directly to group B, where the more vulnerable possibilities are obviously found. But is this reasonable? Note that the expected value we will obtain in point of generality is given by the quantity:

$$.05x(20) + .005y(80) = x + .4y$$

And so, let us suppose, we shall take to maximize this subject to the basic constraints:

$$x \leqslant 10$$
$$y \leqslant 100$$
$$10x + y < 120$$

The result is a neat little linear-programming problem:

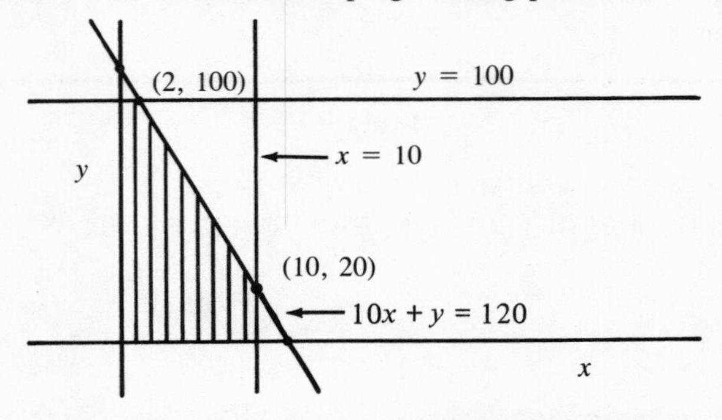

So $x + .4y$ is maximized with a value of 42 at (2, 100) and we are to work through all 100 of the group B hypotheses and two others (randomly chosen?) group A hypotheses. The high-generality hypotheses definitely wind up in the preferred position here.

But, of course, had the case been somewhat different, the result would differ too. Thus, if the group A hypotheses were a bit more general and the group B ones a bit less so (say 40 and 50, respectively), the case would be reversed. For then the expected generality is:

$$.05x(40) + .005y(50) = 2x + .25y$$

And this is maximized at (10, 20), so that the group A hypotheses would now enjoy priority.

The point is that our economically oriented approach is wholly undogmatic regarding generality-preference. We replace Popper's purely logical concern for universality-for-its-own-sake with an economico-methodological concern for universality-relative-to cost. If we take this economic line and do sensible decision making on the basis of the seemingly reasonable economic precept, "Maximize generality subject to the constraints of affordability," then our basic concern is one of cost-benefit analysis, seeking to optimize returns subject to resource-outlays.[127] From this not unreasonable perspective, it becomes altogether secondary whether priority is given to the more or the less general alternatives. The dispute of generality-priority versus specificity-priority now looks rather like a doctrinaire bit of unrealism. Once due heed is paid to the economic aspects of the matter ideological attachment on the basis of "general principles" to highly general (or highly specific) hypotheses becomes a luxury we can no longer afford.

7. Probabilism versus Improbabilism

Let us turn next to the Popperian thesis that the rational agenda of scientific inquiry calls on science to give priority to the less probable (more vulnerable) hypothesis in a given problem-situation.[128] The case stands as follows:

Issue: We are confronted with a series of alternative hypotheses to resolve a given problem. All these are supposed to lie on the same level in point of generality. We are to put their respective tenability to the test.

Question: Do we want to test the hypotheses in order of the

> *probability* or their *improbability*? Are the serious candidates for priority concern the probable or the improbable ones?

Here, then, we confront the issue of a Carnapian probability-preference versus a Popperian improbability-preference in relation to scientific hypotheses.[129]

Consider, for the sake of illustration, the case of three alternative hypotheses that are credited with the following probabilities in the face of the evidence-at-hand:

hypothesis:	H_1	H_2	H_3
probability:	.1	.2	.7

Suppose we are to proceed in a series of pairwise tests of these H_i against one another. Suppose further that the possible outcomes of these paired test comparisons simply have a likelihood determined by the relative weights of their initial probabilities. We now pose the question whether to test the probables against the probables first, or the improbables against the improbables. The supposition just made leads to the following tabulation of test-outcome probabilities.

Test \ Victor	$\Pr(H_i/H_i \text{ vs. } H_j)$		
	H_1	H_2	H_3
H_1 vs. H_2	.33	.66	—
H_1 vs. H_3	.13	—	.87
H_2 vs. H_3	—	.22	.78

There remains, of course, the question of the cost of these comparison tests. We shall make a very simple assumption here, that of the intrinsically plausible principle:

> *Ceteris paribus*, the cost of pairwise tests *decreases* monotonically with the increasing distance between the likelihoods of the competitors involved.

We shall thus suppose, for simplicity of illustration, the simple proportionality principle:[130]

$\text{cost}(H_i \text{ vs. } H_j) \approx 1 - \Delta(H_i, H_j)$ and so $= k[1 - \Delta(H_i, H_j)]$ where
$$\Delta(H_i, H_j) = |\, \text{pr}(H_i/H_i \text{ vs. } H_j) - \text{pr}(H_j/H_i \text{ vs. } H_j)|$$

We thus arrive at a tabulation of test costs:

Test	*Cost*
H_1 vs. H_2	$.67k$
H_1 vs. H_3	$.26k$
H_2 vs. H_3	$.44k$

1. The Popperian strategy: work on unlikeliest first:

$$H_1 \text{ vs. } H_2 \begin{cases} \xrightarrow{.33} H_1 \text{ vs. } H_3 \\ \xrightarrow{.66} H_2 \text{ vs. } H_3 \end{cases}$$

Expected cost:

$$.33(.67k + .26k) + .66(.67k + .44k) = 1.04k$$

2. The orthodox strategy: work on likeliest first:

$$H_2 \text{ vs. } H_3 \begin{cases} \xrightarrow{.22} H_2 \text{ vs. } H_1 \\ \xrightarrow{.78} H_3 \text{ vs. } H_1 \end{cases}$$

Expected cost:

$$.22(.44k + .67k) + .78(.44k + .26k) = .78k$$

In this example, the orthodox strategy of first testing the more probable alternatives against one another proves superior on an expected-cost basis, but there is nothing inevitable about this. In other cases the situation could eventuate differently.

Consider the preceding example subject to a change of probabilities:

hypothesis:	H_1	H_2	H_3
probability:	.1	.4	.5

The conditions of the problem are now changed to:

	H_1	H_2	H_3
H_1 vs. H_2	.20	.80	—
H_1 vs. H_3	.17	—	.83
H_2 vs. H_3	—	.44	.56

Test	*Cost*
H_1 vs. H_2	$.4k$
H_1 vs. H_3	$.34k$
H_2 vs. H_3	$.88k$

1. The Popperian strategy:

$$H_1 \text{ vs. } H_2 \begin{cases} \xrightarrow{.2} H_1 \text{ vs. } H_3 \\ \xrightarrow{.8} H_2 \text{ vs. } H_3 \end{cases}$$

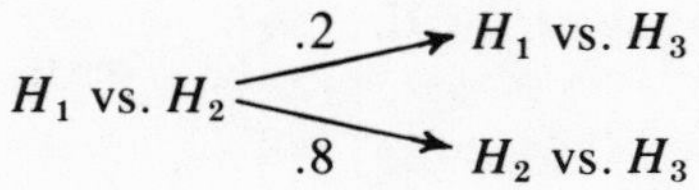

Expected cost:

$$.2(.4k + .34k) + .8(.4k + .88k) = 1.17k$$

2. The orthodox strategy:

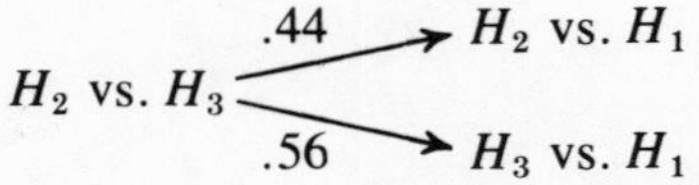

$$H_2 \text{ vs. } H_3 \begin{cases} \xrightarrow{.44} H_2 \text{ vs. } H_1 \\ \xrightarrow{.56} H_3 \text{ vs. } H_1 \end{cases}$$

Expected cost:

$$.44(.88k + .4k) + .56(.88k + .34k) = 1.24k$$

In this case, the Popperian strategy of giving priority to the least probable hypothesis proves to be the most cost-effective.

We thus see that the economic approach is eminently "pragmatic." It does not accord general superiority either to a strategy of probability-priority or improbability-priority. It is neutral on this particular ideological issue, and lets the chips fall where they may—or rather, where economic factors of the Peircean sort indicate they should.

Note that throughout our discussion, what is at issue is not the economic advantageousness of selecting a certain hypothesis for *adoption* (or "rational acceptance") but rather the quintessentially abductive issue of selecting a certain hypothesis for *testing*. The distinction between researchworthy hypotheses and credible theories is crucial here. Our attention is directed to the design of a *research program* and not the much later issue of theory-acceptance.

8. Simplicity

It is recognized by theoreticians on all sides that simplicity must play a prominent part in the methodology of science. Agreement is as widespread as any to be found on philosophical matters that simple hypotheses are to be preferred both for plausibility and credibility in inductive contexts. On the other hand, when we push the issue back to first principles and press the question of the rationale of this preference for simplicity, we meet with a note of indecisiveness. One recent commentator has remarked, "The whole discussion of simplicity has been curiously inconclusive. Not only has there been no growing body of agreement concerning the measurement of simplicity, but there has been no agreement concerning . . . the precise role that simplicity should play in the acceptance of scientific hypotheses."[131] The literature of the problem is replete with indications of such uneasiness. The principle seems less congenial nowadays to tough-minded philosophical analysts than it did to old-school metaphysicians with a penchant toward quasitheological ideas like the simplicity of nature.

It is difficult to justify simplicity-preference on grounds either of informativeness or of falsifiability.[132] It is easy to do so on grounds of economy. If we claim that a phenomenon depends not just on certain distances and weights and sizes but also, say, on temperature and magnetic forces, then we must operate a more complex testing apparatus and contrive to take readings over this enlarged range of physical parameters. Or again, in a certain curve-fitting case, compare the thesis that the resultant function is linear

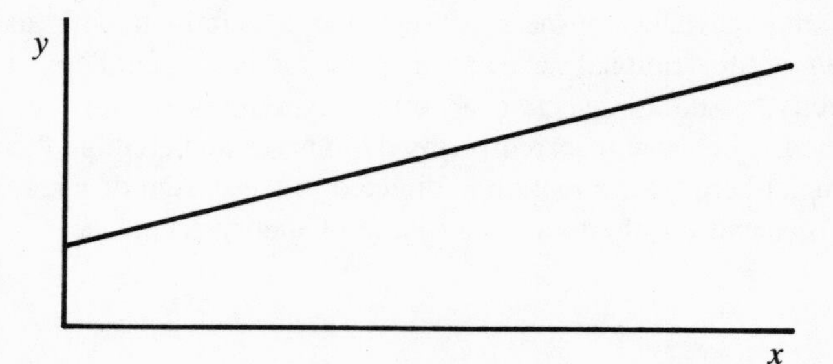

with the thesis that it is linear up to a point and sinusoidally wave-like thereafter:

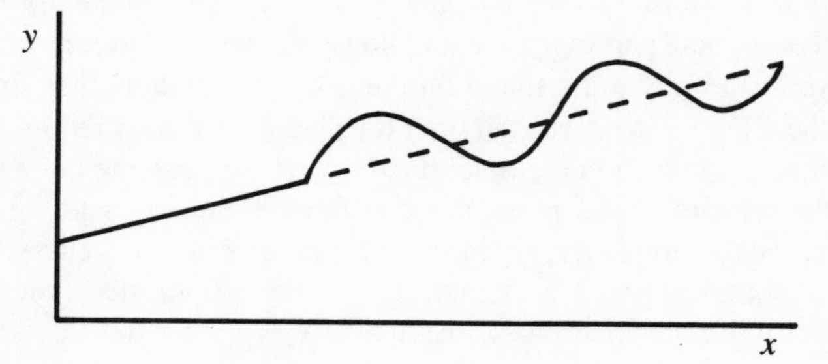

Now think of writing a computer program to check whether empirically determined point-coordinates fit the specified function. This is clearly a vastly less complex, and so more economical, process in the linear case than in its more convoluted congener. Comparable considerations of operative economy attach to simplicity on the side not just of *substantiation*, but also of *utilization*, making simpler alternatives more advantageous to select for adoption as well as for testing. The concerns of economy provide a straightforward and reasonable basis for a policy of simplicity.

On this approach, no claim is made (tacitly or otherwise) for any sort of ontological linkage between simplicity and (probable) truth. Simplicity-preference is based on the strictly practical ground that simple hypotheses are operationally cheaper, the most advantageous for us to put to use. There is no need for recourse to a substantive (or constitutive) postulate of the simplicity of nature; it suffices to have recourse to a regulative (or practical) precept of economy of means. This amounts to a principled defense of simplicity that in fact rests on practical considerations.

Peirce himself offered essentially this sort of justification for simplicity-preference. He maintained that the combination of relatively good probability of correctness with a relatively low cost of testing gave simple hypotheses a preferred status, since the expected-value calculations of the economy of research indicate that (other things being at all equal) simple hypotheses can be put into operation more advantageously than others.[133]

9. Conclusion

Enough has now been said to indicate the main point emerging from these various illustrations. The conception of the economy of research in the actual conduct of inductive inquiry provides an instrument of considerable power. Even in a very rudimentary form, the economic perspective emphasized by Peirce can straightforwardly resolve some of the key disputes in the recent theory of inductive reasoning, including problems involving Carnap's requirement of total evidence, Hempel's paradox of the ravens, Goodman's grue paradox, the concept of simplicity-preference, and some of the chief issues controverted between Popper and his opponents.

It is interesting and significant to note that a common thread runs through many of the major controversial points in the recent theory of inductive reasoning. Time and again, the theoreticians put before us a sort of story where common sense says something quite different, and there arises a great controversy over issues such as grue versus blue, or black ravens versus white shoes, issues where the whole tradition of ordinary practice tells us that there is no practical occasion for worry. It now becomes clear why clever theoreticians here encounter perplexities to which the ordinary practitioner seems immune. They have all too commonly left out of consideration one of the very central factors of this (and any other) enterprise, the economic element. In consequence, a gap opens between the theoreticians' ideas based on the abstract logic of things and the ordinary practitioner's sense of natural procedure. Taking account of the economic aspect of inductive practice can neatly close this gap in many cases. Peirce's vision was clearer here than that of most of his successors.

A great deal remains to be said both about the economy of research itself and about Peirce's ideas on the subject, but I trust that what has been put forward here is enough to persuade the reader that, in giving this subject a central place in his theory of inductive inquiry, Peirce was well ahead of his contemporaries, and indeed of ours as well.[134]

Having reproached contemporary methodologists of science with insufficient attention to the economy of research as conceived by Peirce, it is in order for this concluding section to do somewhat fuller justice in this regard.

For Peirce the "economy of research" encompasses the economics of inductive inquiry as a whole. It specifically includes:

1. The economics of *abduction*, including the economic aspects of concept-development and hypothesis-projection. This would include, for example, such questions as "Do we proceed with inquiry on the basis of the hypotheses we have in hand or should we continue further with the work of hypothesis-proliferation?" and "Of the (all too vast) mass of hypotheses we have 'available' which ones should we bother to test and which can be safely ignored?"
2. The economics of *retroduction* (hypothesis-testing), including (a) the design of economical programs for programs and efficient testing-strategies and (b) the design of efficient tactics in evidence gathering including, for example, such questions as "Do we decide *now* between the hypotheses under consideration or do we continue with evidence-gathering?"

The present-day situation is this: Problem-area 1, the economics of hypothesizing, is an entirely untouched domain; while area 2, the economics of hypothesis-testing, is somewhat active, but remains in an extremely rudimentary state of development. The prime focus of recent attention has been point (2b), and the main result of these discussions has been the distinctly unexciting conclusion that it pays to acquire new information in hypothesis-testing when this can be done free of cost (or cost-irrelevantly).[135] Most recent discussions in this general area, however, do not deal with the economics of inquiry itself at all, but only with the economics of hypothesis acceptance—with maximizing the expected utility of the result of inquiry.[136] Above all, the crucial realm of the economics of abduction—how we should conduct our hypothesis-formulating efforts, and, in particular, just how much labor we should expend in the abductive stage of inquiry and how our resources may be expended most efficiently here—is a problem-area that is altogether untouched.[137] Peirce's ideas still have not been fully assimilated to present-day approaches.[138]

Notes

1. Peirce's Theory of the Self-correctiveness of Science

1. CP, 2.769 [c. 1905]; compare ibid., 1.67. (Regarding notes of this form, see p. 117.)

2. Compare also CP, 2755ff., 2.770ff., 2.781, 5.145, 5.574ff., 7.111, and 7.207.

3. CP, 7.78 [c. 1902].

4. See CP, 7.110ff.

5. CP, 2.756. As Peirce says, this rests on the "Pooh-pooh argument" which "proceeds from the premiss that the reasoner has no evidence of the existence of any fact of a given description and concludes that there never was, is not, and never will be any such thing" (CP, 7.111 [1903]).

6. For a particularly clear statement see CP, 2.758: "[Quantitative induction] presumes that the value of the proportion, among the *S*'s of the sample, of those that are *P*, probably approximates, within a certain limit of approximation, to the value of the real probability [of an *S*'s being a *P*]."

7. CP, 2.770 [c. 1905].

8. For the extensive literature on "the straight rule," see Wesley C. Salmon, *The Foundations of Scientific Inference* (Pittsburgh, 1967).

9. The first sentence of this passage is simply incorrect. Peirce does not say that all forms of induction are self-corrective, but that every branch of science is (CP, 5.579, 5.582).

10. John Lenz draws a veil of charity over the matter, observing that Peirce's "Remarks on the self-correcting nature of the broader form of induction are extremely hard to comprehend" (Lenz 1964, p. 152, n. 2). C. Y. Cheng objects in a similar vein: "Peirce does not make it clear what the self-correcting process of induction means . . ." (Cheng 1969, p. 67).

11. To be sure, Duhemian considerations introduce complications into the

determination of whether or not a seemingly "unsuccessful application" is in fact to be counted as such.

12. Note that, given Peirce's theory of probability, this ratio represents in the long run the *probability* that an application of the theory will be successful. To say that this probability is 1 is to say that the theory is true. And the closer to 1 its probability of applicative success, the "nearer to truth" is the theory. For Peirce's theory of probability, see Madden 1964, and Burks 1964. Peirce thus never really rejects his own earlier attempts to provide a wholly probabilistic legitimation of induction with reference to the law of large numbers. See Cheng 1966 and 1967 and compare Delaney 1973.

13. In *The Logic of Scientific Discovery* (New York, 1959), K. R. Popper said, "A theory is a tool which we test by applying it, and which we judge as to its fitness by the results of its applications" (p. 108). If we de-instrumentalize this statement by changing "tool" to "thesis" and "fitness" to "truth," we obtain a claim that would not be out of place in Peirce's writings. (In its original formulation it smacks of a somewhat later pragmatist, William James.)

14. Hans Reichenbach, *Theory of Probability* (Berkeley, 1949), p. 475.

15. See especially CP, 6.40.

16. The use of success ratios enables one to assess whether one's theorizing efforts manage to approach the truth more closely without presupposing that it is necessary in determining this actually to know what the truth is. This meets the frequent objection that "Peirce's thesis that induction must lead to truth in a self-correcting process must presuppose some particular criteria for determining the truth in practice, if induction is valid in his sense at all" (Cheng 1967, p. 27).

17. See Ronald Fischer, *Statistical Methods and Scientific Inference* (Edinburgh, 1956).

18. For the historical background, see Laudan 1973, pp. 275–306.

19. Joseph Priestley, *The History and Present State of Electricity* (London, 1767), p. 381.

20. Georges Le Sage, "Quelques opuscules relatifs à la méthode," published posthumously in Pierre Prévost. *Essais de philosophie,* vol. 2 (Paris, 1804), pp. 253–335.

21. The following line of objection:

> [T]he method of quantitative inductions provides no machinery whatever for satisfying . . . [a] necessary condition for a self-corrective method; *given that* an hypothesis has been refuted [in retroductive testing], qualitative induction specifies no technique for discovering an alternative *H′* which is (or is likely to be) closer to the truth than the refuted *H*. Nor does it even provide a criterion for determining

> whether an alternative H' is closer to the truth than H. Peirce, in short, gives no persuasive arguments to establishing that quantitative induction is . . . self-corrective. (Laudan 1973, p. 290)

22. Peirce's own statements are sometimes careless on this issue, because he failed (in earlier writings) to distinguish carefully enough between theory-selection ("abduction" or "hypothesis") and theory-confirmation ("induction" in general, including specifically its abductive part). Both involve hypothesis elimination, the former on grounds of plausibility appraisals, the latter on the basis of testing. Thus Peirce writes late in his life to Paul Carus: "in almost everything I printed before the beginning of this century I more or less mixed up hypothesis and induction. . . " (CP, 8.227 [c. 1910]).

23. It is in this synoptic sense that we must construe the oft-cited passage: "So it appears that this marvelous self-correcting property of Reason, which Hegel made so much of, belongs to every sort of science, although it appears as essential, intrinsic, and inevitable only in the highest type of reasoning, which is induction" (CP, 5.579 [1898]).

24. Peirce does not think it possible that the scientific method might lead to any ultimately irreparable error. He holds that inquiry is *predestined* ultimately to yield the truth even in the hands of imperfect practitioners as long as the scientific method is strictly adhered to:

> [I]f you take the most pigheaded and passionate of men who has sworn by all the gods that he never will allow himself to believe the earth is round, and give him time enough, and cram that time with experience in the pertinent sphere . . . he will surely come to and rest in the truth about the form of the earth. Such is the infallibility of science. (CP, 7.78 [c. 1902])

As we shall see in the next chapter, however, the self-monitoring of science by quantitative induction is in fact unable to prevent the possibility that scientific inquiry might indeed end in an effectively irreparable error.

25. In working out the argument of this chapter, I have profited by exchanges with William C. House.

2. Peirce on Scientific Progress and the Completability of Science

26. If "life and mental vigor were to be indefinitely prolonged," as he puts it in one passage (CP, 8.41 [c. 1885]). Again: "Truth is that concordance of an abstract statement with the ideal limit towards which endless investigation would tend to bring scientific belief" (CP, 5.565 [1901]).

27. We must not be parochial and conceive of "the ongoing community of inquirers" as necessarily limited to human beings:

> We may take it as certain that the human race will ultimately be extripated. . . . But, on the other hand, we may take it as certain that other intellectual races exist on other planets,—if not of our solar system, then of others; and also that innumerable new intellectual races have yet to be developed; so that on the whole, it may be regarded as most certain that intellectual life in the universe will never finally cease. (CP, 8.43 [c. 1885])

28. CP, 5.407 [1878]. In the essay "What Pragmatism Is" (CP, 5.416–34), Peirce proposes "to define the 'truth' as that in a belief to which belief would lead if it were to tend indefinitely toward absolute fixity."

29. Occasionally Peirce says things like, "the validity of induction depends merely on there being any reality" (CP, 5.349 [1869]). But this is careless; strictly construed, his theory is that the dependence is reversed: reality of the only sort that can concern the pragmaticist, knowable reality, depends on the validity of induction, its self-corrective capacity ultimately to uncover whatever statistical stabilities may be there. Peirce himself came to recognize the carelessness at issue and ultimately rejected his earlier view that scientific reasoning requires a reality postulate. (See the important 1884 paper, "Design and Chance" [Ms. 875]. I am grateful to Dr. H. William Davenport for providing a photocopy of it.) Cf. note 44 below.

30. Consider again J. S. Mill's contention, advanced in his famous essay "On Liberty," that the competition among schools of thought is akin to that between biological varieties: the rivalry among ideas for acceptance amounts to a struggle for existence (that is, perpetuation or continued existence), a struggle in which those beliefs that are the fittest, those that represent "the truth," must finally prevail. A parallel is drawn between survival of beliefs within an intellectual community and biologicial survival. Peirce's position is an inverted version of Mill's: Mill says that if a maintained thesis is true, then it must ultimately prevail; Peirce says that if a thesis ultimately prevails, then it must be true.

31. To be sure, even Peirce himself had his moments of queasiness here:

> [W]hen we discuss a vexed question, we *hope* that there is some ascertainable truth about it, and that the discussion is not to go on forever and to no purpose. A transcendentalist would claim that it is an indispensable "presupposition" that there is an ascertainable true answer to every intelligible question. I used to talk like that, myself; for when I was a babe in philosophy my bottle was filled from the udders of Kant. But by this time I have come to want something more substantial. (2.113 [c. 1902])

Admittedly, Peirce sometimes has his doubts here as elsewhere; he is a prismatic thinker who sometime toys with a plurality of discordant positions. Thus he concludes a most interesting passage with the contention, "we cannot know that there *is* any truth concerning any given questions," but then goes on to add, "But practically, we know that questions do generally get settled in time, when they come to be scientifically investigated; and that is practically and pragmatically enough" (CP, 5.494 [1907]; compare 7.569 [1892]). Reconciliation can always be found in Peirce's fallibilism and the idea that further inquiry could always show our best-founded old beliefs to have been, not altogether wrong, but inexact. It appears that until about 1880 Peirce held that "there is an ascertainable true answer to every intelligible question" (compare CP, 2.113 [c. 1902]), no doubt following W. K. Clifford's contention that "to every reasonable question there is an intelligible answer which either we or posterity may know by the exercise of scientific thought." (See "On the Aims and Instruments of Scientific Thought" [1872], reprinted in L. Stephen and F. Pollack, eds., *W. K. Clifford: The Ethics of Belief and Other Essays,* [London 1898], pp. 25–26.) Thereafter, Peirce seems to have begun to have doubts about the flat-out truth of this thesis as a constitutive principle, for "we cannot be quite sure that the community will ever settle down to an unalterable conclusion upon any given question" (CP, 6.610 [1893], compare 5.494 [1907]). But he continued to hold that (1) constitutively speaking, it is *almost* always true, and (2) it must be maintained as a regulative principle (Cf, MS 875 [1884]; CP, 8.43ff. [1885]). Compare note 70 below.

Peirce thus maintained a considered, consistent position in his later years (after 1880):

> The problem whether a given question will ever get answered or not is not so simple; the number of questions asked is constantly increasing, and the capacity for answering them is also on the increase. If the rate of the latter increase is greater than that of the [former] the probability is unity that any given question will be answered; otherwise the probability is *zero*. Considerations too long to be explained here lead me to think that the former state of things is the actual one. In that case, there is but an infinitesimal proportion of questions which do not get answered, although the multitude of unanswered questions is forever on the increase. . . . But I will admit (if the reader thinks the admission has any meaning, and is not an empty proposition) that some finite number of questions, we can never know which ones, will escape getting answered forever. (CP, 8.43 [c. 1885])

32. For Peirce, science is effectively a latter-day surrogate—a functional equivalent, as it were—for the medieval philosopher's conception of the

"mind of God": idealized long-run science is infallible with respect to nature (thesis [1]), and is moreover also omniscient with respect to nature (thesis [2]). ("I should say that God's omniscience, humanly conceived, consists in the fact that knowledge in its development leaves no question unanswered" [CP, 8.44].) The real truth about factual matters, on such a view, simply coincides with what science maintains to be so in the long run, and so genuine knowledge can be construed as final irreversible opinion.

33. Though Peirce is a pragmatist, he does *not* (with William James) take truth as determinable through pragmatic efficacy. Thus Bertrand Russell is, as regards Peirce, quite wrong in saying that the crucial difference between the pragmatist's theory and his own is that he defines truth in terms of the *causes* and they in terms of the *effects* of beliefs (*An Inquiry into Meaning and Truth* [London, 1940] [Penguin reprint 1963], see p. 308.). Peirce's conception of truth is realistic and correspondistic; true beliefs are "satisfactory" in virtue of their consonance with reality and are not in danger of subsequent overthrow once arrived at.

34. D. A. Bromley et al., *Physics in Perspective* Student Ed. (Washington, D.C., 1973; National Research Council/National Academy of Science Publication), p. 26.

35. CP, 5.565 [1901].

36. CP, 1.116 [c. 1896].

37. Quoted in *Physics Today* 21 (1968): 56. See also Charles Weiner, "Who Said It First?", *Physics Today* 21 (1968):9. Compare the somewhat ampler statement in A. A. Michelson, *Light Waves and Their Uses* (Chicago, 1961):

> The more important fundamental laws and facts of physical science have all been discovered, and these are now so firmly established that the possibility of their ever being supplanted in consequence of new discoveries is exceedingly remote. Nevertheless, it has been found that there are apparent exceptions to most of these laws, and this is particularly true when the observations are pushed to a limit, i.e., whenever the circumstances of experiment are such that extreme cases can be examined. . . . Many other instances might be cited, but these will suffice to justify the statement that "our future discoveries must be looked for in the sixth place of decimals." It follows that every means which facilitates accuracy in measurement is a possible factor in a future discovery. . . . (pp. 23–24)

The "eminent physicist" who made this claim is presumed to be Lord Kelvin (compare the exchange of letters in *Science* 172 (1971):111, as well as p. 52 of Badash's paper cited in note 39). I have not succeeded in verifying this attribution; it does not square with Kelvin's oft-repeated views. This perception of him as a prime exponent of late nineteenth-century confidence in the completeness of science does him a grave injustice.

38. T. C. Mendenhall, *A Century of Electricity* (Boston and New York, 1887; revised 1890), p. 223.

39. For an interesting (but very incomplete) study of this idea, see Lawrence Badash, "The Completeness of Nineteenth-Century Science," *Isis* 63 (1972):48–58.

40. Raymond T. Birge, "Physics and Physicists of the Past Fifty Years," *Physics Today* 9 (1956):20ff.

41. Max Planck, *Vorträge und Erinnerungen*, 5th ed. (Stuttgart, 1949), p. 169.

42. Thus for Ostwald the "progress in discovery we experience anew from day to day . . . affords a guarantee that in the course of time one wish after another will be satisfied and one possibility after another will be realized, so that science will approach the ideal of omnipotence with rapid steps" (Wilhelm Ostwald, *Die Wissenschaft* [Leipzig, 1911], p. 47). Compare the following passage from a well-known textbook: "All branches of theoretical physics, with the exception of electricity and magnetism, can be regarded at the present state of science as concluded, that is, only immaterial changes occur in them from year to year" (Charles Emerson Curry, *Theory of Electricity and Magnetism* [New York and London, 1897], p. 1; I owe this reference to Martin Curd.).

43. Letter to Daniel Coit Gilman [1878], printed in Wiener and Young 1952, pp. 365ff. (see p. 366).

44. CP, 8.12 [1871]. Peirce goes on to say that "any truth more perfect than this destined conclusion, any reality more absolute than what is thought in it, is a fiction of metaphysics."

45. Bertrand Russell in P. A. Schilpp, ed., *The Philosophy of John Dewey* (New York, 1951), p. 23.

46. Though not, to be sure, earlier on. There are many passages where Peirce stresses the revolutionary nature of scientific development preceeding its ultimate evolutionary phase (see especially CP, 1.108–109 [c. 1896.]).

47. Peirce also had moments of doubt about this. In one passage he writes:

> [T]he largest proportion of all among those who derive their ideas of physical science from reading popular books . . . the Spencers, the Youmanses, and the like, seem to be possessed with the idea that science has got the universe pretty well ciphered down to a fine point; while the Faradays and Newtons seem to themselves like children who have picked up a few pretty pebbles upon the ocean beach. (CP, 5.65 [1903]; compare 6.610 [1893])

48. William Whewell was perhaps the first philosopher of science to reject the theory of the cumulativity of progress:

> It would, however, be a great mistake to suppose that the hypotheses, among which our choice thus lies, are constructed by an enumeration of obvious cases, or by a wanton alteration of relations which occur in

> some first hypothesis. It may, indeed, sometimes happen that the proposition which is finally established is such as may be formed, by some slight alteration, from those which are justly rejected. . . . But it more frequently happens that new truths are brought into view by the application of new Ideas, not by new modifications of old ones. (*Novum Organon Renovatum* [3rd ed., London, 1858], bk. 2, chap. 4, sect. 9: "New Hypotheses Not Mere Modifications of Old Ones")

49. See Lawrence Badash's paper, cited at note 39, above.

50. *The Coming of the Golden Age* (Garden City, 1969), pp. 111–112.

51. Note the similarity of the Kuhnian picture of "normal science." See Thomas Kuhn, *The Structure of Scientific Revolutions* (Chicago, 1962).

52. For a fuller development of some of the issues of this section, see the author's *Scientific Progress* (Oxford, 1977).

53. This shibboleth of the contemporary philosophy of science is not all that new. Already at the turn of the century, Sir Michael Foster had written:

> The path [of progress in science] may not be always a straight line; there may be swerving to this side and to that; ideas may seem to return again and again to the same point of the intellectual compass; but it will always be found that they have reached a higher level—they have moved, not in a circle, but in a spiral. Moreover, science is not fashioned as is a house, by putting brick to brick, that which is once put remaining as it was put to the end. The growth of science is that of a living being. As in the embryo, phase follows phase, and each member or body puts on in succession different appearances, though all the while the same member, so a scientific conception of one age seems to differ from that of a following age. . . . ("The Growth of Science in the Nineteenth Century," *Annual Report of the Smithsonian Institution For 1899* [Washington, 1901], pp. 163–183 [as reprinted from Foster's 1899 presidential address to the British Association for the Advancement of Science]; see p. 175)

54. A detailed exposition and defense of this view of cognitive progress is given in the author's *Methodological Pragmatism* (Oxford, 1976).

55. The apple analogy is perhaps misleading in one respect. The product of production certainly need not be something discrete, like "an apple" or "a discovery." Cost-escalation says that it costs more to produce a unit amount starting at $t + \Delta t$ than at t. Yield-diminution says that it takes longer to produce a unit amount starting at $t + \Delta t$ than at t. The one is clearly possible without the other: they coincide only if production-times and production-costs are proportional, which would be rather a special case.

56. These data are drawn from Derek de Solla Price, *Little Science, Big Science* (New York, 1963).

57. Price, pp. 92–93. Before the 1970s, expenditure on science in the U.S., measured in constant dollars, had been doubling every 5–6 years.

58. Price, p. 39.

59. For details on this and related issues see chapter 4 of the author's *Scientific Progress* (Oxford, 1977).

60. Raymond H. Ewell, "The Role of Research in Economic Growth," *Chemical and Engineering News* 33 (1955):2980–2985.

61. It is worth noting for the sake of comparison that for more than a century now the total U.S. federal budget, its nondefense subtotal, and the aggregate budgets of all federal agencies concerned with the environmental sciences (Bureau of Mines. Weather Bureau, Army Map Service, etc.) have all grown at a uniform annual rate of 9 percent. (See H. W. Menard, *Science: Growth and Change* [Cambridge, Mass., 1971], p. 188.)

62. The U.S. Federal government's expenditures for research and development came to 13.8 billions in 1965 (60% for defense, 33% for space, and 7% for the rest); Ewell did not reckon with the cost implications of the space race. For these data see the NSF biennial series "An Analysis of Federal R & D Funding by Function."

63. Data from William George, *The Scientist in Action* (New York, 1938).

64. Du Pont's outlays for research stood at $1 million annually during World War I (1915–1918), $6 million in 1930, $38 million in 1950, and $96 million in 1960. (Data from Fritz Machlup, *The Production and Distribution of Knowledge in the United States* [Princeton, 1962], pp. 158–159; and see pp. 159–160 for the relevant data on a larger scale.) Overall expenditures for scientific research and its technological development (R & D) in the U.S. stood at $\$.11 \times 10^9$ in 1920, $\$.13 \times 10^9$ in 1930, $\$.38 \times 10^9$ in 1940, $\$2.9 \times 10^9$ in 1950, $\$5.1 \times 10^9$ in 1953–1954, $\$10.0 \times 10^9$ in 1957–1958, $\$11.1 \times 10^9$ in 1958–1959, and about $\$14.0 \times 10^9$ in 1960–1961 (ibid., pp. 155 and 187). Machlup thinks it not unreasonable to suppose that no other industry or economic activity in the U.S.A. has grown as fast as R & D (ibid., p. 155).

65. "Impact of Large-Scale Science on the United States," *Science* 134 (21 July 1961):161–164 (see p. 161). Weinberg further writes:

> The other main contender [apart from space exploration] for the position of Number One Event in the Scientific Olympics is high-energy physics. It, too, is wonderfully expensive (the Stanford linear accelerator is expected to cost $\$100 \times 10^6$), and we may expect to spend $\$400 \times 10^6$ per year on this area of research by 1970. (Ibid., p. 164)

66. *Vorträge und Erinnerungen,* 5th ed. (Stuttgart, 1949), p. 376.

67. For if economic constraints limit our access to data (by restricting our interactions with nature), they are bound to impede the reach of our theories as well.

68. The somewhat telegraphic discussion of this section is developed more fully in the author's *Scientific Progress* (Oxford, 1977).

69. Further details about this project of Peirce's are given in the last chapter.

70. But he goes on to say, "As far as all ordinary and practical questions go I insist upon this axiom as much as ever,—as much as anybody can do" (pp. 7–8). And in the text partly quoted above, Peirce holds that "there is but an infinitesimal proportion of questions which do not get answered" (CP, 8.43 [c. 1885]). Until about 1880, his commitment to this "axiom" was total and unqualified (compare footnote 31 above), but in his later writings Peirce became more cautious. In particular he acknowledged that even if we knew on general principles that an irrevocable answer would ultimately be reached, we could not ever know with assurance at any particular juncture that the answer we then had in hand was irrevocable. Still, he goes on to insist that only if a irrevocable answer can be reached will the thing at issue have a definite reality or being:

> [T]o say that a thing *is* is to say that in the upshot of intellectual progress it will attain a permanent status in the realm of ideas. Now, as no experiential question can be answered with absolute certainty, so we never can have reason to think any given idea will either become unshakably established or be forever exploded. But to say that neither of these two events will come to pass definitively is to say that the object has an imperfect and qualified existence. (CP, 7.569 [c. 1892])

In one place Peirce describes the view that all scientific questions are scientifically answerable as merely "a cheerful hope" (CP, 5.407 [1878]). This is rather a comedown for a "fundamental axiom of logic"!

71. Or again:

> There may be questions concerning which the pendulum of opinion never would cease to oscillate, however favorable circumstances may be. But if so, those questions are *ipso facto* not *real* questions, that is to say, are questions to which there is no true answer to be given. (CP, 5.460 [1905])

One Peircean example of a question science might conceivably find to be unsettleable is that of the geometric structure of physical space as Euclidean or Riemannian or Lobatchavskian. (See his *Math. Papers,* ed. Carolyn Eisele, and note 82 below.)

72. CP, 2.113 [c. 1902]. Compare note 31 above.

73. It is perhaps clear enough how Peirce would want this to be construed. Reality, for him, is inherently cognizable and what science can come to know are the permanent regularities experientially accessible over the long

term. Thus scientific knowledge cannot reach wherever there is utter irregularity and disorder, and the sectors of nature where these prevail are *eo ipso* "unreal." Science proceeds at the level of the general; it does not deal with irregularity; that part of nature lies outside its purview. (The position is strongly reminiscent of Plato and Aristotle.) There is thus a sector of nature, a sector of surdity and chaos, that lies beyond scientific rationalization and is thus unreal. Peirce seems committed the view that this sector is ever narrowing in the course of cosmic evolution, so that in the end the real (that is, the rational and the scientifically cognizable and rationalizable) and the natural (existential, actual) will increasingly coincide. But this long-run conformation of cognizable reality with natural actuality is certainly rather a metaphysical hope than an argued doctrine. Yet without this hope the thesis that "science will discover whatever is knowable" comes down to the arid, nearly trivial claim that science will discover in the long run everything that can be discovered scientifically in the long run. And so without this hope Peirce's basic contention that science will discover the real truth of things (in the theoretical long run) is trivialized, in that truth is understood simply as what corresponds to the intelligible part of Nature (i.e., "reality"). Knowledge of the rest of nature is excluded by definition: what we cannot have knowledge of just *is* this extra-real "Nature." Nevertheless, Peirce's hope about how far knowledge can reach cannot be what his overall purposes require him to hold it to be: a substantive, constitutive truth that describes the way things in fact are. Peirce sometimes leaves this hope at the level of a regulative assumption—"we shall have to be governed by it practically" (i.e., in actual practice, as in the above quotation). To do its job in Peirce's argumentation, the "cheerful hope" cannot rest at the level of a regulative precept concerning our comportment as inquirers operating within nature; it must be maintained as a constitutive thesis about how nature itself is constituted. And this aspect of the matter reflects a serious defect.

74. "[A]ll unknowable reality is nonsense" (CP, 8.43; p. 46 [c. 1885]).

75. The following passage—perhaps Peirce's fullest treatment of the issue—is typical:

> Let us suppose, then, for the sake of argument, that some questions eventually get settled, and that some others, indistinguishable from the former by any marks, never do. In that case, I should say that that conception of reality was rather a faulty one, for while there is a real so far as a question that will get settled goes, there is none for a question that will never be settled; for an unknowable reality is nonsense. The nonidealistic reader will start at this last assertion; but consider the matter from a practical point of view. You say that real things are manifested by their effects. True . . . and if nothing is ever settled about the matter, it will be because the phenomena do not

> consistently point to any theory; and in that case there is a want of that "uniformity of nature" (to use a popular but very loose expression) which constitutes reality, and makes it differ from a dream. In that way, if we think that some questions are never going to get settled, we ought to admit that our conception of nature as absolutely real is only partially correct. Still, we shall have to be governed by it practically; because there is nothing to distinguish the unanswerable questions from the answerable ones, so that investigation will have to proceed as if all were answerable. In ordinary life, no matter how much we believe in questions ultimately getting answered, we shall always put aside an innumerable throng of them as beyond our powers. We shall not in our day seek to know whether the centre of the sun is distant from that of the earth by an odd or an even number of miles on the average; we shall act as if neither man nor God could ever ascertain it. There is, however, an economy of thought, in assuming that it is an answerable question. From this practical and economical point of view, it really makes no difference whether or not all questions are actually answered, by man or by God, so long as we are satisfied that investigation has a universal tendency toward the settlement of opinion. . . . (CP, 8.43 [c. 1885])

The passage is worth pondering. While it is not a monument of cogent reasoning, it manages to compress all of Peirce's characteristic doctrines on this subject into a compact statement.

76. The line of thought at issue in this paragraph is more fully developed in the author's *Methodological Pragmatism* (Oxford, 1976).

3. Peirce on Abduction, Plausibility and the Efficiency of Scientific Inquiry

77. A useful exposition of the fundamental ideas of Peirce's theory of knowledge is given in William H. Davis, *Peirce's Epistemology* (The Hague, 1972).

78. CP, 6.524 [c. 1901].

79. CP, 5.181 [1903]. In another passage Peirce writes:

> We cannot go so far as to say that high human intelligence is more often right than wrong in its guesses; but we can say that, after due analysis, and unswerved by prepossessions, it has been, and no doubt will be, not very many times more likely to be wrong than right. (CP, 7.220 [c. 1901])

80. CP, 7.220 [c. 1901].

81. Their evolutionary origin serves to delimit the area over which our hypothesis-selective instincts will be relatively trustworthy:

> If we subject the hypothesis, that the human mind has such a power [of successful abduction] in some degree, to inductive tests, we find that there are two classes of subjects in regard to which such an instinctive scent for the truth seems to be proved. One of these is in regard to the general modes of action of mechanical forces, including the doctrine of geometry; the other is in regard to the ways in which human beings and some quadrupeds think and feel. In fact, the two great branches of human science, physics and psychics, are but developments of that guessing-instinct under the corrective action of induction. (CP, 6.531 [c. 1901])

82. As science progresses more and more, and moves ever further into regions of physical condition (temperature, pressure, size, etc.) increasingly remote from the setting of ordinary human life, our instinctive sense of the plausible becomes less and less serviceable:

> [A]s we penetrate further and further from the surface of nature, instinct ceases to give any decided answers; and if it did, there would no longer be any reason to suppose its answers approximated to the truth. (CP, 7.508 [c. 1898])

And again:

> As we advance further and further into science, the aid we can derive from the natural light of reason becomes, no doubt less, and less; but still science will cease to progress if ever we shall reach the point where there is no longer an infinite saving of expense in experimentation to be effected by care that our hypotheses are such as naturally recommend themselves to the mind. . . . (CP, 7.220 [c. 1901])

Thus already in 1891 C. S. Peirce had suggested that while the axioms of Euclidean geometry had served man so well that evolution had rendered them "expressions of our inborn conception of space" still "that affords not the slightest reason for supposing them exact," and indeed the universe may well be non-Euclidean (CP, 6.29 [1891]).

83. Cf. CP 2.753 [1883]; 5.172 [1903]; 5.591 [1903]; 6.476 [1908]; 6.531 [c. 1901]; and especially the following passage:

> But that we have made solid gains in knowledge is indisputable; and, moreover, the history of science proves that when the phenomena were properly analyzed . . . it has seldom been necessary to try more than two or three hypotheses made by clear genius before the right one was found. (CP, 7.220 [1901])

84. There is another fundamental abduction whose defense is of such a this-or-nothing sort. For Peirce insisted perhaps more emphatically than any other philosopher of science, that the "uniformity of nature" (which he construed as the complex of two principles: that the whole is as the observed sample, and that similarity in discerned respects is an indicator of similarity in others) is not a law of nature but a condition of factual knowledge in general, a part of the "formal conditions of all knowledge" (CP, 7.138 [c. 1902]; cf. 7.581). Regarding this sector of Peirce's thought see Delaney 1973, pp. 438–448.

85. Dunn 1972, p. 37.

86. For a fuller development of these lines of thought, see the author's *Methodological Pragmatism* (Oxford, 1977).

87. The following passage elegantly depicts Peirce's position:

> It is a great mistake to suppose that the mind of the active scientist is filled with propositions which, if not proved beyond all reasonable cavil, are at least extremely probable. On the contrary, he entertains hypotheses which are almost wildly incredible, and treats them with respect for the time being. Why does he do this? Simply because any scientific proposition whatever is always liable to be refuted and dropped at short notice. A hypothesis is something which looks as if it might be true and were true, and which is capable of verification or refutation by comparison with facts. The best hypothesis, in the sense of the one most recommending itself to the inquirer, is the one which can be the most readily refuted if it is false. This far outweighs the trifling merit of being likely. For after all, what is a *likely* hypothesis? It is one which falls in with our preconceived ideas. But these may be wrong. Their errors are just what the scientific man is out gunning for more particularly. But if a hypothesis can quickly and easily be cleared away so as to go toward leaving the field free for the main struggle, this is an immense advantage. (CP, 1.120 [c. 1896])

And again,

> It is a very grave mistake to attach much importance to the antecedent likelihood of hypotheses, except in extreme cases; because likelihoods are most merely subjective, and have so little real value, that considering the remarkable opportunities which they will cause us to miss, in the long run attention to them does not pay. Every hypothesis should be put to the test by forcing it to make verifiable predictions. (CP, 5.599 [1903])

Note that these remarkable passages were written by Peirce around 1900, and not by Karl Popper a generation later.

88. This resort to the mechanism of trial and error as a basic model for the

scientific method is not confined to Popperians. Stephen C. Pepper, for example, also maintained that "the inductive methods of experimental science are essentially systematized trial and error," and has based on this idea a rather sophisticated Darwinian model of knowledge (as well as of value). See his book, *The Sources of Value* (Berkeley and Los Angeles, 1958), whence the preceding quote (from p. 106). The stress on *blindness* in lieu of *randomness* is eminently well advised. One knows from information theory that, for example, if the game of Twenty Questions were played perfectly—so that each yes or no yields the contestant one bit of information—then twenty bits would suffice to identify 2^{20} candidate-objects (that is, one out of more than a million). But this prospect requires that there must be nothing random about these questions—they require vast "background knowledge" for the careful partitioning of the information-space at every stage. For "blind" inquiry to prove efficient, it must pertain only to one small speck in the cognitive field. Efficiency here demands the demonically shrewd deployment of an "inductive talent" of the kind that Popper finds unpalatable.

89. In Popper's book there is a stress on mutation-reminiscent randomness or near randomness ("more or less accidental trial-and-error gambits," "almost random or cloud-like trial-and-error movements") that is revoked in Popper's Schilpp essay (*The Philosophy of Karl Popper*, ed. by P. A. Schilpp [2 vols., La Salle, 1974]). "I regard this idea of the 'blindness' of the trials in a trial-and-error movement as an important step beyond the mistake idea of random trials" (p. 1062).

90. *Objective Knowledge* (Oxford, 1972), p. 28.

91. Here there is a substantial disanalogy with the evolutionary case. Darwin did not need to include unicorns in the purview of the theory and explain their nonexistence by some process akin to an account for the extinction of dinosaurs.

92. Compare Peirce's very similar view:

> [A]ll that logic warrants [with respect to the success of inquiry] is a *hope*, and not a belief. . . . [W]hen we discuss a vexed question, we *hope* that there is some ascertainable truth about it, and that the discussion is not to go on forever and to no purpose. (CP, 2.113 [c. 1902])

Or again:

> [A]ll the followers of science are animated by a cheerful hope that the processes of investigation, if only pushed far enough, will give one certain solution to each question to which they apply it. (CP, 5.407 [1878])

But despite the surface similarity, Peirce's position is quite different from Popper's. On Peirce's view, we can and do have empirical evidence for the convergence of inquiry: that is, all that logic can do is to offer a hope, but science can go further and show that this hope is warranted; it can come to be "retrospectively revalidated," as I have put it elsewhere (*Methodological Pragmatism* [Oxford, 1977]). For Popper, the hope must forever remain unjustified.

93. This "too little time" complaint is reminiscent of the objections once offered by William Thomson, Lord Kelvin, against Darwinian evolution. In his presidential address to the British Association in 1871, he complained that its mechanism of natural selection was "too like the Laputan method for making books, and that it did not sufficiently take into account a continually guiding and controlling intelligence. . . ." However inappropriate this objection may be deemed in the case of *biological* evolution, the situation is quite otherwise in the case of *cognitive* evolution. For an interesting account of the time-availability dispute between physicists on the one side and biologists and geologists on the other, see Stephen G. Brush, "Thermodynamics and History" in *The Graduate Journal* 2 (1969):477–565.

94. CP, 5.172–173 [1903]. Compare the following passage:

> Nature is a far vaster and less clearly arranged repertory of facts than a census report, and if man had not come to it with special aptitudes for guessing right, it may well be doubted whether in the ten or twenty thousand years that they may have existed their greatest mind would have attained the amount of knowledge which is actually possessed by the lowest idiot. (CP, 2.753 [1883])

95. This point warrants emphasis. Most writers on induction who hold that man has (or develops) inductive skills do so to assure that our conjectures have a nontrivial *a priori* probability for the purpose of Bayesian argumentation. Peirce sees that this is also needed to rationalize *the relatively rapid rate of scientific progress*. On this aspect of Peirce's thought compare Sharpe, 1970.

96. The issues of this problem-area have perhaps been pursued more effectively by Herbert A. Simon than by any other cognitive theoretician. See his essay "Does Scientific Discovery Have a Logic?" *Philosophy of Science* 40 (1973):471–480, where further references to his work are given. One key summary runs thus: "The more difficult and novel the problem, the greater is likely to be the amount of trial and error required to find a solution. At the same time, the trial and error is not completely random or blind; it is, in fact, highly selective" (*The Sciences of the Artificial* [Cambridge, Mass., 1969], p. 95). Exploration of the computer simulation of the processes of human learning and discovery brings clearly to light the operation of an essentially regulative/methodological heuristic. It is based on principles (such as the priority of "similarity"-augmenting transformations in

problem-solving) which *qua* theses are clearly false (are heuristic "fictions" in the sense of Vaihinger), but which prove methodologically effective.

97. The methodological approach can thus lay claims to resolving the issue perceptively posed by D. T. Campbell:

> Popper has, in fact, disparaged the common belief in "chance" discoveries in science as partaking of the inductivist belief in directly learning from experience. . . . [T]hat issue, and the more general problem of spelling out in detail the way in which a natural selection of scientific *theories* is compatible with a dogmatic blind-variation-and-selective-retention epistemology remain high priority tasks for the future. (P. A. Schilpp, ed., *The Philosophy of Karl Popper*, 2 vols. [La Salle, 1974], p. 436)

The present theory provides a natural basis for combining a natural selection process at the level of *theories* with an epistemology of blind-variation-and-selective-retention at the level of *methods*.

98. On the broader aspects of Peirce's views on evolution see Goudge 1964.

99. The idea of such a developing methodology of discovery (or "logic of discovery") is itself altogether in sympathy with Peirce's ideas. (See CP, 2.108 [c. 1902].) For some modern developments along these lines, which would surely be congenial to him, see Norwood R. Hanson, *Patterns of Discovery* (Cambridge, 1958), and his influential 1961 paper "Is There a Logic of Discovery?" in H. Feigl and G. Maxwell, eds., *Currect Issues in the Philosophy of Science*, vol. 1 (New York, 1961). The work of Herbert A. Simon is an important development in this area: "Thinking by Computers" and "Scientific Discovery and the Psychology of Problem Solving" in R. G. Colodny, ed., *Mind and Cosmos* (Pittsburgh, 1966); Alan Newell and H. A. Simon, *Computer Simulation and Human Thinking* (New York, 1961).

100. Some of the lines of thought of this chapter are developed more fully in the author's *Methodological Pragmatism* (Oxford, 1976).

4. Peirce and the Economy of Research

101. Peirce's main paper on the subject is an 1878 essay of this title. (CP, 7.139–57; but for the date see 5.601) Compare also CP, 7.158–61 [1902], 7.220 [1901] and 1.122–125 [c. 1896]. Peirce returned to the subject at length in a detailed proposal put before the Carnegie Institution (unsuccessfully) in 1902 (CP, 7.158–161).

Peirce tended to give Ernst Mach credit as a precursor or congener (quite erroneously), beginning his opuscule on the Economy of Research with a bow in his direction: "Dr. Ernst Mach, who has one of the best faults a philosopher can have, that of riding his horse to death, does just this with his principle of Economy in science" (CP, 1.122; compare also CP, 5.601). To be sure, Mach had given a widely circulated 1882 lecture on "The Economical Nature of Physical Inquiry" (see his *Popular Scientific Lectures* [Chicago, 1895, pp. 186–213); but this postdated Peirce's own 1876–1879 investigations. In any case, Mach's démarche came down to pushing his "convenient abbreviation" theory of physical laws as an (imperfect) compression of a mass of facts into a compact formula: "[T]o save the labor of instruction and of acquisition, concise, abridged description is sought. This is really all that natural laws are" (*Popular Scientific Lectures*, tr. T. J. McCormack [Chicago, 1895], p. 193). The resort to "economy" was for Mach simply a convenient peg on which to hinge his doctrine of laws; he accorded economic considerations no formative role in the framework of scientific inquiry or of inductive reasoning itself. Peirce did, however, lean on Mach's ideas in formulating his own conception of induction as "a species of 'reduction of the manifold to unity' " (CP, 5.275 [1893]) insofar as it aims at the lawful synthesis of numerous facts.

102. "Abduction furnishes all our ideas concerning real things . . . but is mere conjecture, without probative force. . . . Induction gives us the only approach to certainty concerning the real that we can have" (CP, 8.209 [c. 1905]).

103. "[T]he number of possible hypotheses concerning the truth or falsity of which we really know nothing, or next to nothing, may be very great. In questions of physics there is sometimes an infinite multitude of such possible hypotheses. The question of economy is clearly a very grave one" (CP, 6.530 [c. 1901]).

104. Some of the lessons for hypothesis-testing that reuse extracts from considerations of the economy of research are set out at CP, 7.223–231.

105. Thus Peirce wrote:

> Perhaps we might conceive the strength, or urgency, of a hypothesis as measured by the amount of wealth, in time, thought, money, etc., that we ought to have at our disposal before it would be worth while to take up that hypothesis for examination. In that case it would be a quantity dependent upon many factors. Thus a strong instinctive inclination towards it must be allowed to be a favouring circumstance, and a disinclination an unfavourable one. Yet the fact that it would throw a great light upon many things, if it were established, would be in its favour; and the more surprising and unexpected it would be to find it true, the more light it would generally throw. The expense which the examination of it would involve must be one of the main factors of its urgency. (CP, 2.780 [1902])

In another passage CP, 7.223 [c. 1901], Peirce offers the following taxonomy:

Economical Considerations
regarding the "experiential character" of hypotheses
Cheapness
Intrinsic Value
—Naturalness
—Likelihood
Relation of Hypotheses
—Caution
—Breadth
—Incomplexity

106. Compare the interesting discussion at CP, 1.85–86 [1898].
107. Hence the important place in inquiry which Peirce assigned to the principle of Occam's Razor; see CP, 4.35.
108. The abductive talent of the preceding chapter plays a crucial role here:

> Is he [the scientist] not free to examine what theories he likes? The answer is that it is a question of economy. If he examines all the foolish theories he might imagine, he never will (short of a miracle) light upon the true one. Indeed, even with the most rational procedure, he never would do so, were there not an affinity between his ideas and nature's ways. However, if there be any attainable truth, as he hopes, it is plain that the only way in which it is to be attained is by trying the hypotheses which seem reasonable. . . . (CP, 2.776 [1902])

On the other hand, Peirce does not rate the value of mere likelihood (antecedent personal probability, as it were) very highly in itself, for it must be conditioned by considerations of the design of experiments. In one very Popperian passage he writes:

> The best hypothesis, in the sense of the one most recommending itself to the inquirer, is the one which can be the most readily refuted if it is false. This far outweighs the trifling merit of being likely. For after all, what is a *likely* hypothesis? It is one which falls in with our preconceived ideas. But these may be wrong. Their errors are just what the scientific man is out gunning for more particularly. But if a hypothesis can quickly and easily be cleared away so as to go toward leaving the field free for the main struggle, this is an immense advantage. (CP, 1.120 [1903])

In another place he derogates likelihoods:

> But experience must be our chart in economical navigation; and ex-

> perience shows that likelihoods are treacherous guides. Nothing has caused so much waste of time and means, in all sorts of researches, as inquirers' becoming so wedded to certain likelihoods as to forget all the other factors of the economy of research; so that, unless it be very solidly grounded, likelihood is far better disregarded, or nearly so; and even when it seems solidly grounded, it should be proceeded upon with a cautious tread, with an eye to other considerations, and a recollection of the disasters it has caused. (CP, 7.220 [c. 1901])

109. Besides its role in hypothesis-testing, considerations of economy of research also enter into the costs of *acting* on an hypothesis. (CF. CP, 7.142) We shall not here treat this separable issue as separate. Successful application is, after all, a crucial aspect of testing.

110. Thus Peirce wrote:

> This [economic] value [of our knowledge on a given subject] increases with the fullness and precision of the information, but plainly it increases slower and slower as the knowledge becomes fuller and more precise. The cost of the information also increases with its fullness and accuracy, and increases faster and faster the more accurate and full it is. It therefore *may* be the case that it does not pay to get *any* information on a given subject; but, at any rate, it *must* be true that it does not pay (in any given state of science) to push the investigation beyond a certain point in fullness or precision. (CP, 1.122 [c. 1896]; CP, 1.84)

Peirce was perhaps the first to draw attention to the rising cost of experimental work as science matures in the course of its historical development (Cf. CP, 7.144). On this issue see the author's *Scientific Progress* (Oxford, 1977).

111. See his printed 1879 paper on "Economy of Research" (CP, 7.139ff.).

112. He referred to it as that "which really is in all cases the leading consideration in Abduction, which is the question of Economy—Economy of money, time, thought, and energy" (CP, 5.600 [1903]).

113. About a third of this current material appears in the quarterly *Transactions of the C. S. Peirce Society*, and the rest elsewhere. For the bibliography regarding Peirce see "A Draft of a Bibliography of Writings about C. S. Peirce" by Max Fisch in *Studies in the Philosophy of Charles Sanders Peirce* ed. E. C. Moore and R. S. Robin (Amherst, 1964), pp. 486–511 (ca. 400 entries) and the two supplements thereto in *Transactions of the Charles S. Peirce Society,* vol. 2 (1966), with ca. 100 entries and vol. 10 (1974), with ca. 350 entries.

114. Peirce lived from 1839–1914. His fortieth year, the inception of the Greek *akmē* or Roman *floruit*, thus began in 1878.

115. Cushen 1967 is only a seeming exception; actually it is a reprint of Peirce's own essay. Murray G. Murphey's detailed and excellent *The Development of Peirce's Philosophy* (Cambridge, Mass., 1961) has no index entry for "economy of research." The 1968 doctoral dissertation by Raymond M. Herbenick and his 1970 article based on it are in part relevant to the topic. K. T. Fann's 1970 monograph *Peirce's Theory of Abduction* has a brief section (pp. 47–51) on economy of research, consisting only of bare summarizing of some Peircean passages without much elaboration or analysis. Sharpe 1970 contains a few incidental observations on the economy of research. By contrast, Peirce's work in theoretical economics has begun to be duly appreciated, as witness W. J. Baumol and S. M. Goldfeld, eds., *Precursors in Mathematical Economics* (London, 1968). See also Carolyn Eisele's paper "Charles S. Peirce and the Mathematics of Economics" (*Proceedings of the Thirteenth International Congress of the History of Science* [Moscow, 1974], vol. 5 [*History of Mathematics*], pp. 171ff.), which gives a brief description of Peirce's project.

116. It is, moreover, highly topical in extraphilosophical contexts. "Research planning"—operations research as applied to scientific research and development—is an active component in current technological-administrative studies. Economists have also become interested in the macrolevel aspects of the issue. See Keith Norris and John Vaizey, *The Economics of Research and Technology* (London, 1973).

117. *Logical Foundations of Probability* (Chicago, 1950; 2d ed., 1962), p. 211. For a most useful survey of the literature on inductive inference see Henry E. Kyburg, Jr., "Recent Work in Inductive Logic," *American Philosophical Quarterly* 1 (1964):249–287. A particularly illuminating discussion of the total evidence requirement is given in Wesley C. Salmon, *Statistical Exploration and Statistical Relevance* (Pittsburgh, 1969).

118. In these instances, the Carnapian requirement is tantamount to Reichenbach's rule of basing statistical probabilities on the narrowest reference class for which statistics can be had. See Hans Reichenbach, *The Theory of Probability* (Berkeley and Los Angeles, 1949).

119. It is very important in this context to realize that economic considerations do not necessarily militate toward narrow reference classes. In a statistics gathering day and age, information may be more readily and cheaply accessible on a nationally aggregated basis than at a more localized and highly disaggregated level. We have the interesting (if paradoxical) situation that information at a high level of generality and large-scale aggregation is usually more cheaply and easily come by than more relevant, case-specific information.

120. C. G. Hempel, "A Note on the Paradoxes of Confirmation," *Mind* 55 (1946):79–82. See also Rudolf Carnap, *Logical Foundations of Probability*, pp. 223–224.

121. Note, however, the prospect that in some circumstances the numbers involved could be such that case would be otherwise.

122. The situation would be quite otherwise if the "big change" came about in a gradual and unnoticed way, as a slow shift with no rude shocks to our memory of how things used to look, or else with a gradual readjustment, so that the recollection of discrepancies would fade away in the manner of an image-inversion experiment. For then we would continue to make our inductive projections on the old basis; for example, grass would still be described as *green*, even though it "really" looks blue.

123. For a survey of objections and positions see Henry E. Kyburg, Jr., "Recent Work on Inductive Logic," who summarizes his discussion with the observation "The problem of finding some way of distinguishing between sensible predicates like 'blue' and 'green' and the outlandish ones suggested by Goodman, Barker, and others, is surely one of the most important problems to come out of recent discussions of inductive logic" (p. 266).

124. This idea of utilizing ostension as a device for tackling Goodman's paradox is originally due to Wesley C. Salmon, "On Vindicating Induction," in *Induction: Some Current Issues* ed. H. E. Kyburg, Jr., and E. Nagel (Middletown, 1963), pp. 27–41.

125. As Israel Scheffler—perhaps the ablest of Goodman's expositors and defenders—has remarked: "The most natural objection to Goodman's new approach is that it provides no explanation of entrenchment itself" ("Inductive Inference: A New Approach," *Science* 27 (January 24, 1958):177–181).

126. *The Logic of Scientific Discovery* (New York, 1959), p. 273.

127. Looking on theories as intellectual instrumentalities (for explanation, prediction, etc.), one can apply to them the usual economic considerations of many sided versatility vs. case-specific power that applies to tools in general. On the economic aspect of this latter issue, compare the suggested discussion, "The Law of Diminishing Returns in Tools," in George Kingsley Zipf, *Human Behavior and the Principle of Least Effort* (Boston, 1949), pp. 66ff. (and cf. also pp. 182ff.).

128. Peirce himself tends to share this Popperian predilection:

> The best hypothesis, in the sense of the one most recommending itself to the inquirer, is the one which can be the most readily refuted if it is false. This far outweighs the trifling merit of being likely. For after all, what is a *likely* hypothesis? It is one which falls in with our preconceived ideas. But these may be wrong. Their errors are just what the scientific man is out gunning for more particularly. But if a hypothesis can quickly and easily be cleared away so as to go toward leaving the field free for the main struggle, this is an immense advantage. (CP, 1.120 [c. 1896])

129. Popper put the issue as follows:

> Thus if we aim, in science, at a high informative content—if the growth of knowledge means that we know more, that we know *a* and *b* rather than *a* alone, and that the content of our theories thus increases—then we have to admit that we also aim at a low probability . . . and since a low probability means a high probability of being falsified, it follows that a high degree of falsifiability, or refutability, or testability, is one of the aims of science. (*Conjectures and Refutations* [London, 1963], p. 219)

Note that on this point too Popper walks (as he often does) in the footsteps of Peirce:

> Experiment is very expensive business, in money, in time, and in thought; so that it will be a saving of expense, to begin with that positive prediction from the hypothesis [under investigation] which seems least likely to be verified. For a single experiment may absolutely refute the most valuable of hypotheses, while a hypothesis must be a trifling one indeed if a single experiment could establish it. (CP, 7.206 [c. 1901], compare 1.120)

130. Nothing fundamental to the structure of the discussion would be altered by taking the cost to be governed by some other reasonable principle, such as proportionality to the ratio of the largest of pr(H_i/H_i vs. H_j) and pr(H_j/H_i vs. H_j) to the smallest of these two.

131. Henry E. Kyburg, Jr., "Recent Work . . ." pp. 267–268.

132. It is dogma among the Popperians that a simple theory is more readily falsified than a complex one (compare K. R. Popper, *The Logic of Scientific Discovery* [London, 1969], chap. 7). But this is true only if the complex theory is *ad hoc* designed specifically to eliminate the falsifiers of the simple one, that is, if the complex theory is a bad theory. It is certainly not true in general. (Compare Stephen F. Barker, "On Simplicity in Empirical Hypotheses," *Philosophy of Science* 28 (1961):162–171; reprinted in M. H. Foster and M. L. Martin, eds., *Probability, Confirmation and Simplicity* (New York, 1966), an anthology that contains several very good papers on the topic.

133. For Peirce's argumentation see CP, 6.530–532.

134. Recent inductive theorists who have proposed cognitive decision models concerned with accepting hypotheses have generally ignored problems that arise in designing experiments or collecting data. Statisticians have done much better at facing up to these issues. Peirce had the good fortune to be as much one as the other.

135. See J. J. Good, "On the Principle of Total Evidence," *The British Journal for the Philosophy of Science*, 17 (1966–1967):319–321; 340–342; Risto Hilpinen, "On the Information Provided by Observations" in *Informa-*

tion and Inference, ed. J. Hintikka and P. Suppes (Dordrecht, 1970), pp. 97–122; W. K. Goosens, "The Logic of Experimentation" (Stanford, Ph.D. dissertation, 1970). To some extent the "cost of search" approach to hypothesis selection envisaged in Howard Raiffa, *Decision Analysis* (Reading, Mass., 1968) establishes a point of contact between items (1) and (2).

136. These discussions go back to Rudolf Carnap's treatise *The Logical Foundations of Probability* (Chicago, 1950). Since Carnap's day these problems have been widely canvassed.

137. In a widely discussed 1953 paper ("The Scientist Qua Scientist Makes Value Judgments," *Philosophy of Science* 20 [1953]:1–6, reprinted in *Readings in the Philosophy of Science*, ed. B. Brody [Englewood Cliffs, 1969], pp. 540–546), Richard Rudner has posed the second-order research-design question: How long must inquiry be conducted to assure the (contextually appropriate) probabilification of a certain probability-assignment? Rudner's discussion implies (without treating) the problem of balancing the cost of continued research against the risks of mistakes due to accepting imperfect information. This sort of question is decidedly a Peircean issue and Rudner deserves credit for helping to keep it on the agenda. But I think it not unfair to say that, having posed this sort of issue, Rudner does not do anything with it. In this regard, Peirce not only maintains priority but primacy as well.

138. In writing this chapter I have been able to profit from the comments of my colleagues John Haugeland and Gerald J. Massey on an earlier version, written for publication in *Philosophy of Science* 43 (1976):71–98. In revising it for inclusion here I have benefited from suggestions by William Davenport and Teddy Seidenfeld.

References

CP: *Collected Papers of Charles Sanders Peirce,* ed. Charles Hartshorne, Paul Weiss, and Arthur Burks, 8 vols. Cambridge, Mass.: Harvard University Press, 1931–1958.

TCSPS: *Transactions of the Charles S. Peirce Society*

The ensuing list registers secondary material directly relevant to the themes of this book. For a full-scale bibliography see: Max Fisch, "A Draft of a Bibliography of Writings about C. S. Peirce," in *Studies in the Philosophy of Charles Sanders Peirce,* ed. E. C. Moore and R. S. Robin. Amherst: University of Massachusetts Press, 1964. First supplement to the preceding in TCSPS 2 (1966); second supplement in TCSPS 10 (1974).

Almeder 1975: Robert Almeder, "The Epistemological Realism of Charles Peirce," TCSPS 11 (1975):3–17.

Ayer 1968: A. J. Ayer, *The Origins of Pragmatism: C. S. Peirce and William James*. San Francisco, 1968.

Ayim 1974: Maryann Ayim, "Retroduction = The Rational Instinct," TCSPS 10 (1974):34–43.

Burks 1946: Arthur W. Burks, "Peirce's Theory of Abduction," *Philosophy of Science* 13 (1946):301–306.

Burks 1964: Arthur W. Burks, "Peirce's Two Theories of Probability," in Moore and Robin 1964, pp. 141–150.

Cheng 1966: Chung-Ying Cheng, "Peirce's Probabilistic Theory of Inductive Validity," TCSPS 2 (1966):86–112.

Cheng 1967: Chung-Ying Cheng, "Charles Peirce's Arguments for the Non-Probabilistic Validation of Induction," TCSPS 3 (1967):24–39.

Cheng 1969: Chung-Ying Cheng, *Peirce's and Lewis' Theories of Induction*. The Hague, 1969.

Cushen 1967: W. E. Cushen, "C. S. Peirce on Benefit-Cost Analysis of Scientific Activity," *Operations Research* 14 (1967):641ff.

Delaney 1973: C. F. Delaney, "Peirce on Induction and the Uniformity of Nature," *The Philosophical Forum* 4 (1973):438–448.

Fann 1970: K. T. Fann, *Peirce's Theory of Abduction*. The Hague, 1970.

Fitzgerald 1968: John J. Fitzgerald, "Peirce's Theory of Inquiry," TCSPS 4 (1968):130–143.

Frankfurt 1958: Harry G. Frankfurt, "Peirce's Notion of Abduction," *The Journal of Philosophy* 55 (1958):593–597.

Goudge 1950: Thomas A. Goudge, *The Thought of Charles Sanders Peirce*. Toronto, 1950.

Goudge 1964: Thomas A. Goudge, "Peirce's Evolutionism After Half a Century," in Moore and Robin 1964.

Herbenick 1970: R. M. Herbenick, "Peirce on Systems Theory," TCSPS 6 (1970):84–98.

Laudan 1973: Laurens Laudan, "Peirce and the Trivialization of the Self-Correcting Thesis," in *Foundations of Scientific Method in the 19th Century*, ed. R. Giere. Bloomington, 1973.

Lenz 1964: John Lenz, "Induction as Self-Corrective," in Moore and Robin 1964, pp. 151–162.

Madden 1964: Edward Madden, "Peirce on Probability," in Moore and Robin 1964, pp. 122–140.

Meyers 1971: Robert G. Meyers, "Truth and Ultimate Belief in Peirce," *International Philosophical Quarterly* 11 (1971):87–103.

Moore and Robin 1964: Edward C. Moore and R. S. Robin, eds., *Studies in the Philosophy of Charles Sanders Peirce*, second series. Amherst, 1964.

Murphey 1961: Murray G. Murphey, *The Development of Peirce's Philosophy*. Cambridge, Mass., 1961.

Pinkham 1967: Gordon N. Pinkham, "Some Comments on Cheng, Peirce and Inductive Validity," TCSPS 3 (1967):96–107.

Reilly 1970: F. E. Reilly, *Charles Peirce's Theory of Scientific Method*. New York, 1970.

Savan 1964: David Savan, "Peirce's Infallibilism," in Moore and Robin 1964, pp. 151–161.

Sharpe 1970: Robert Sharpe, "Induction, Abduction, and the Evolution of Science," TCSPS 6 (1970):12–33.

Shimony 1970: Abner Shimony, "Scientific Inference," in *The Nature and Function of Scientific Theories*, ed. R. G. Colodny. Pittsburgh, 1970, pp. 79–172.

Wiener and Young 1952: Philip P. Wiener and Frederick H. Young, eds., *Studies in the Philosophy of Charles Sanders Peirce*. Cambridge, Mass., 1952.

von Wright 1965: Georg Henrik von Wright, *The Logical Problem of Induction*, 2nd ed. Oxford, 1965, p. 226.

Index of Names

Aristotle, 103 n73

Badash, Lawrence, 98 n37, 99 n39, 100 n49
Barker, Stephen, 114 n123, 115 n132
Bergson, Henri, 56
Birge, R. T., 24, 25, 99 n40
Bromley, D. A., 98 n34
Brush, Stephen, 108 n93
Burks, Arthur W., 94 n12

Campbell, D. T., 109 n97
Carnap, Rudolf, 73, 74, 89, 113 n120, 116 n136
Carus, Paul, 8, 95 n22
Cheng, C. Y., 93 n10, 94 nn12, 16
Clifford, W. K., 97 n31
Curd, Martin, 99 n42
Curry, Charles Emerson, 99 n42
Cushen, W. E., 113 n115

Darwin, Charles, 107 n91
Davenport, H. William, 97 n29, 116 n138
Davis, William H., 104 n77
Delaney, C. F., 94 n12, 106 n84

Einstein, Albert, 53
Eisele, Carolyn, 113 n115
Ewell, Raymond, 32, 101 n60

Fann, R. T., 113 n115
Fisch, Max, 112 n113
Fischer, Ronald, 13, 94 n17
Foster, Michael, 100 n53

George, William, 101 n63
Gilman, Daniel Coit, 99 n43
Good, J. J., 115 n135
Goodman, Nelson, 76, 77, 78, 79, 89, 114 nn123, 124, 125
Goosens, W. K., 116 n135
Goudge, T. A., 109 n98

Hanson, Norwood R., 109 n99
Hartley, David, 13
Haugeland, John, 116 n138
Hegel, G. F. W., 95 n23

Hempel, C. G., 74, 89, 113 n120
Herbenick, Raymond M., 113 n115
Hilpinen, Risto, 115 n135
House, William C., 95 n25
Hume, David, 79

James, William, 94 n13, 98 n33

Kant, Immanuel, 19, 96 n31
Kelvin, Lord (William Thomson), 98 n37, 108 n93
Kuhn, Thomas, 29, 100 n51
Kyburg, Henry E., Jr., 113 n117, 114 n123, 115 n131

Laudan, Laurens, 5, 94 n18, 95 n21
Lenz, John, 93 n10
Le Sage, Georges, 13, 14, 94 n20

Mach, Ernst, 109–10 n101
Machlup, Fritz, 101 n64
Madden, Edward, 94 n12
Massey, Gerald, 116 n138
Menard, H. W., 101 n61
Mendenhall, T. C., 23, 99 n38
Michelson, A. A., 23, 98 n37
Mill, J. S., 96 n30
Murphey, Murray G., 113 n115

Newell, Alan, 109 n99
Norris, Keith, 113 n116

Ostwald, Wilhelm, 99 n42

Pepper, Stephen C., 107 n88
Planck, Max, 24, 25, 33, 99 n41
Plato, 103 n73
Popper, K. R., 42, 51–57, 81, 83, 89, 94 n13, 106 n87, 107 nn88, 89, 92, 108 n92, 109 n97, 115 nn129, 132
Prévost, Pierre, 94 n20
Price, Derek de Solla, 100 n56, 101 nn57, 58
Priestly, Joseph, 13, 94 n19

Raiffa, Howard, 116 n135
Reichenbach, Hans, 4, 5, 12, 94 n14, 113 n118
Rudner, Richard, 116 n137
Russell, Bertrand, 26, 98 n33, 99 n45

Salmon, Wesley C., 93 n8, 113 n117, 114 n124
Scheffler, Israel, 114 n125
Seidenfeld, Teddy, 116 n138
Sharpe, R. A., 113 n115
Shimony, Abner, 5
Simon, Herbert A., 108 n96, 109 n99
Stent, Gunther S., 28

Thomson, William (Lord Kelvin), 98 n37, 108 n93

Vaihinger, Hans, 109 n96
Vaizey, John, 113 n116
von Jolly, Phillip, 24

Weinberg, Alvin, 32, 101 n65
Weiner, Charles, 98 n37
Whewell, William, 99 n48

Zipf, George Kingsley, 114 n127

Index of Subjects

abduction, 3, 41–51, 63, 65–72, 95 n22, 104 n79, 105 n83, 106 n84, 108 n94, 110 n102, 111–12 n108, 112 n112
asymptotic approximation, 27
autonomy of science, 16

cognitive Darwinism, 52–54, 57–63
cognitive evolution, 105 n81
collective self-correctiveness, 6–7
community of inquirers, 96 n27
completability of science, 19–39
completeness, 35
co-naturing, 49
consensus, long-run, 96 n30, 96–97 n31, 97–98 n32
Copernican inversion, 19
correctness, 34–35
cost-escalation, 30–35, 100 n55
costs of science, 31–33
cumulative-convergence theory of scientific progress, 26–27
curve-fitting, 87–88

Darwinism
  cognitive, 52–54, 57–63
  instinctival, 62–63
  methodological, 61–63
distributive self-correctiveness, 6–7

economics of science, 35–39
economy of research, 46–47, 65–91, 109–10 n101, 110 n103, 110–11 n105, 112 nn109, 110, 112
entrenchment, 79, 114 n125
evolution, 47–50
evolution, cognitive, 105 n81
evolutionary epistemology, 51–63

falsifiability, 114 n128, 114–15 n129, 115 n132
falsification, 80–83
fictions, 108–9 n96
fin de siècle 21–25

generality-preference, 80–83
geographic exploration model, 27–30
Grue Paradox, 76–79, 114 nn122, 123

hope, 56, 107 n92
hypothesis, 95 n22

ideal limits, 95 n26
ignorabimus, 54
improbabilism, 83–87
induction, 1–4, 41–44, 71–72, 95 n22, 96 n29
inductive method, 1–4
inquirers, community of, 96 n27
insight, 46
instinct, 47–51, 61–63, 108 n94
instinctival Darwinism, 62–63
instrumentalism, 94 n13

likelihood, 106 n87
logicality, 62–63
logic of discovery, 108–9 n96, 109 n99
long-run consensus, 96 n30, 96–7 n31, 97–98 n32

methodological Darwinism, 61–63
methodological turn, 61–63

Occam's Razor, 111 n107
operations research, 113 n116

performance-monitoring, 15
plausibility, 41–51, 61–63
"Pooh-pooh argument," 93 n5
pragmatism, 98 n33
predestination, 95 n24
presumptions, 62
probabilism, 83–87
probability, 94 n12
progress, scientific, 19–39, 98 n37, 99 n42, 99–100 n48, 100 n53, 105 n82

qualitative induction, 2–4, 8–13
quantitative (or statistical) induction, 2–4, 93 n6, 94–95 n21, 95 n24
questions in science, 102 n70, 102 n71, 103–4 n73

randomness, 107 n89
rate of scientific progress, 56–61, 108 n95
Raven Paradox, 74–76
realism, scientific, 19–21, 35–39
reality, 9–10, 12, 20–21, 25–26, 37–39, 96 n29, 102–3 n74
research, economy of, 46–47, 65–91, 109–10 n101, 110 n103, 110–11 n105
research program, 80–87
retroduction, 3, 65–72, 110 n103

science
- autonomy of, 16
- completability of, 19–39
- costs of, 31–33
- economics of, 35–39
- self-correctiveness of, 1–17
- questions in, 102 nn70, 71, 103–4 n73

scientific progress, 19–39, 98 n37, 99 n42, 99–100 n48, 100 n53, 105 n82
- cumulative-convergence theory of, 26–27
- rate of, 56–61, 108 n95

scientific questions, 102 nn70, 71, 103–4 n73
scientific realism, 19–21, 35–39
self-correctiveness, 6–7, 13–16, 93 n10, 94 n23
- collective, 6–7
- distributive, 6–7
- of science, 1–17

simplicity, 87–89
statistical methods, 8–13
success ratios, 9–10, 94 n16

tactics of research, 68–69
time-availability, 58–61, 108 n93
total evidence, 73–74
trial and error, 57–61, 106–7 n88, 109 n97
truth, 20–21, 25–26, 34–35, 37–39, 51–52, 54–55, 95 n26, 96 nn28, 30, 96–97 n31, 98 nn32, 33

ultimate completeness, 20
ultimate correctness, 20

visions, 63

yield-diminution, 30–31, 34–35, 100 n55

zero-growth, 33–35

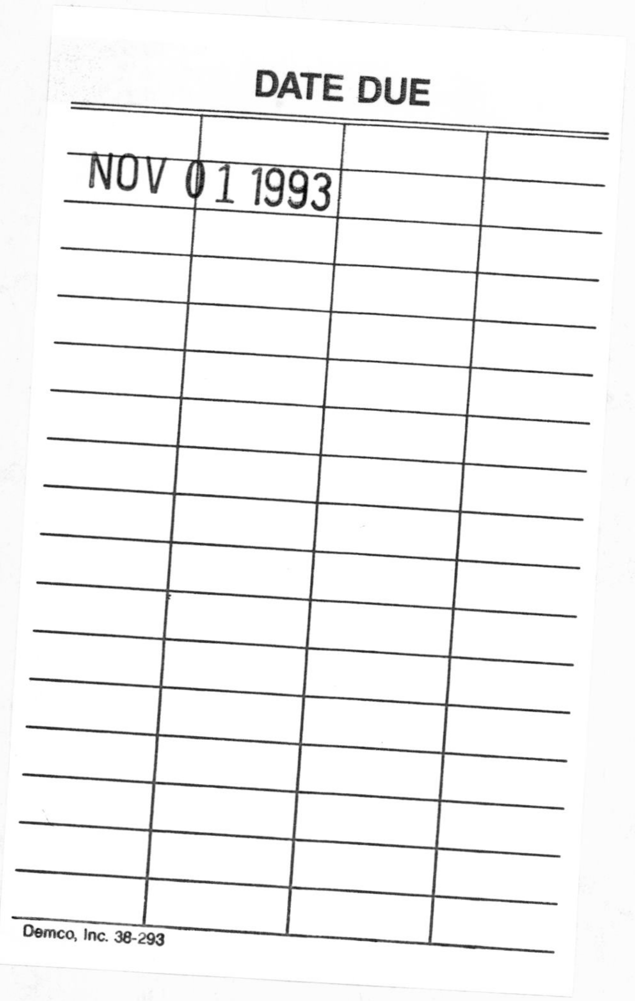
DATE DUE
NOV 01 1993
Demco, Inc. 38-293